10/95

D0031243

MILLIONS and BILLIONS of YEARS AGO

and

# of YEARS AGO: Dating Our Earth and Its Life

by Norman F. Smith

A Venture Book
Franklin Watts
New York · Chicago · London · Toronto · Sydney

For Marj and Bill Devlin

Photographs copyright © : Bruce Con: p. 15; Arizona Office of Tourism: pp. 24 top, 79; U.S. Geological Survey/H.E. Malde: p. 24 bottom; Norman F. Smith: p. 28 top; Jeff Howe: p. 32; American Museum of Natural History/Chris Schuberth, (332982): p. 35; NASA: pp. 70, 99; Metropolitan Museum of Art, Gift of Nelson A. Rockefeller, 1963, (1978.412.6): p. 88; all other photographs courtesy of the author.

Library of Congress Cataloging-in-Publication Data

Smith, Norman F.
Millions and billions of years ago : dating our earth and its life / by Norman F. Smith.
p. cm.—(A Venture book)
Includes bibliographical references and index.
Summary: Discusses the processes by which scientists determine the age of their specimens, such as counting the number of the tree rings, reading the strata, and using chemical analyses and the carbon–14 test. Also looks at recent findings that may tell us how old the universe is.
ISBN 0-531-12533-5
1. Earth—Age—Juvenile literature. 2. Geochronometry—Juvenile literature. [1. Earth—Age. 2. Geochronometry.] I. Title.
QE508.S56 1993
551.7′01—dc20 92-42744 CIP AC

Printed in the United States of America
6 5 4 3 2

CONTENTS

INTRODUCTION
9

1
DATING BY TREE RINGS
11

2
THE AGE OF THE EARTH—
THE OLD VIEW
23

3
LEARNING TO READ STRATA
34

4
CLOCKS IN EARTH ROCKS
50

5
CLOCKS IN SPACE ROCKS
65

6
THE AGE OF THE EARTH—
THE NEW VIEW
74

7
DATING BONES, CHARCOAL,
AND POTTERY
87

8
DATING THE UNIVERSE
97

APPENDIX:
RADIOACTIVITY—A QUICK REVIEW
110

GLOSSARY
117

FOR FURTHER READING
122

INDEX
124

MILLIONS and BILLIONS of YEARS AGO

INTRODUCTION

This earth of ours, on which we find ourselves spinning through space—how has it formed? That question was probably among the first that our ancient ancestors asked when they began to wonder about the world around them. At first, the only answers they could find came from their imaginations, and were recorded in myth, folklore, and religious writings.

More recently, only a few hundred years ago, scientists began to examine the world itself. They began to look more closely at the soil and the rocks, and how they may have changed over the years. They began to ask more searching questions. What processes raised the earth's majestic mountains, carved its rugged canyons, leveled its vast plains?

Because the earth teems with life of all kinds, these early scientists also became curious about life on earth. How did life begin? How did life develop into the vast variety of living things we see today? As scientists began to discover answers to such questions, they found that there was another question for which they urgently needed an answer: *when* did all these things happen?

In order to put the whole story together, early scientists needed a way of dating events in the history of the earth and its living things. As understanding of the processes that formed the earth improved, the search for "clocks" or "calendars" that would reveal how much time the various processes took grew ever more urgent. After many decades, the search was rewarded with the discovery of the first reliable dating methods. The dates that these methods revealed surprised everyone, and changed our whole idea of the universe and the world.

How did we discover ways to look back across millions and billions of years to find out when things happened? As scientists examined the earth more and more closely, they found that changes occur over the years in nearly everything—in rocks, in the oceans, in living things, and in the universe itself. There are, you might say, clocks and calenders hidden everywhere. We only have to find them and learn to read them correctly. Some clocks have given very useful answers. Others have proved unreliable. There are undoubtedly still other clocks that have not yet been discovered.

This book tells the story of how some of these clocks and calenders were discovered and developed into useful tools. The story will take us into history, and into the discoveries made by those sciences that study early life, early man, the earth, and the universe.

Investigating events that occurred in the dim, distant past, long before there were people or any other life on the earth, has been one of science's greatest challenges. Solving the puzzle of not only *what* happened in those times but also *when* it happened has been one of the greatest challenges faced by science.

1 DATING BY TREE RINGS

DATING BY CALENDAR, OR GROWTH CHANGES

This book that you are reading—how long ago was it published? That's an easy question to answer. The date of publication is printed on a page in the front of the book. From this date you can find the exact age of the book in years. You can do this because our society has adopted a calendar that numbers the years and gives us a method of counting the passage of time. Similarly, if someone asks you how old you are, you can answer exactly, because the date of your birth has been recorded and "located" on the calendar.

Even without knowing a person's birth date, we can usually tell roughly how old he or she is. Because people change in many ways as they grow older, we can easily tell whether a person is a baby, an older child, or an adult. We can tell whether the adult is young, middle-aged, or older, and we might even be able to estimate the person's age within a few years. Because every living thing changes as it grows and ages, we can often use its pattern of growth as a clock or calendar to judge how old it is.

TREE RINGS LOG THE YEARS

Some living things have built-in calendars that tell us *exactly* how many years they have lived. Trees, for example, have "growth rings" that form annually as the trees grow. When the growing season begins each year, large thin-walled cells are added below the bark (Figure 1-1). As the end of summer approaches, the added cells become smaller and more thick-walled, until growth ceases altogether in winter. The same pattern is repeated the following year, thus giving the tree a light and dark band each year for as

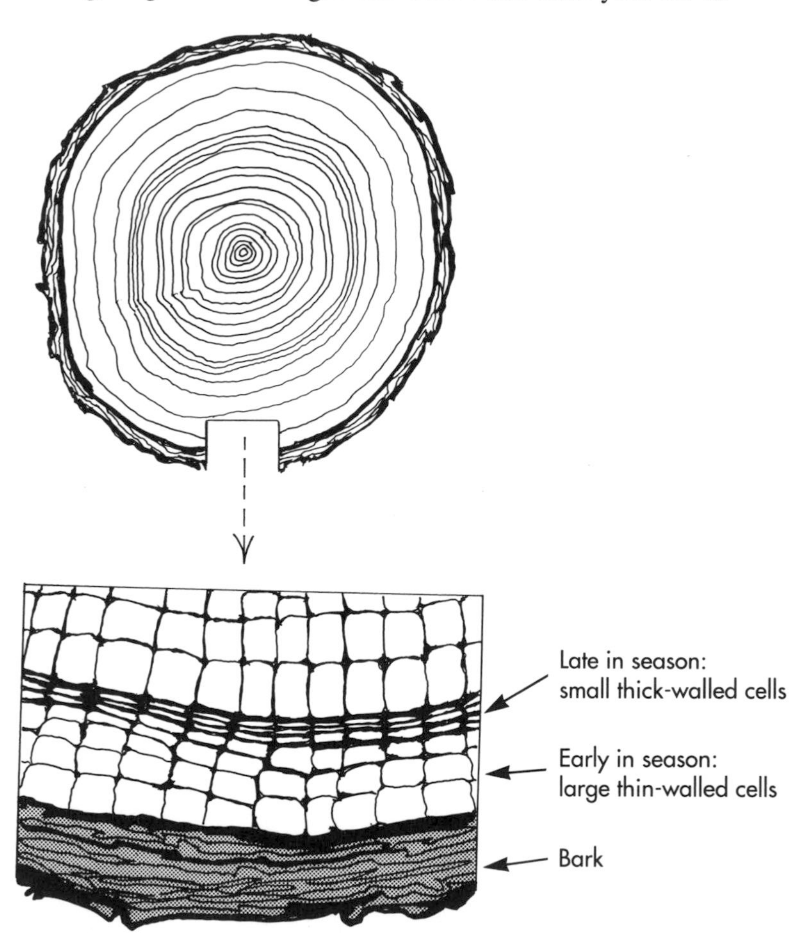

Figure 1-1. The growth process of trees produces annual rings that tell us the number of years that the tree has grown.

long as the tree lives. Counting these growth rings on the end of a cut log or stump will reveal exactly how many years the tree has lived.

You can see growth rings on the end of almost any piece of firewood, lumber, or piling. It is not necessary to cut a tree down to count its rings—a boring tool can be used to drill out a sample, or "core," from the bark to the center of the living tree. The growth rings can be counted from that core (Figure 1-2).

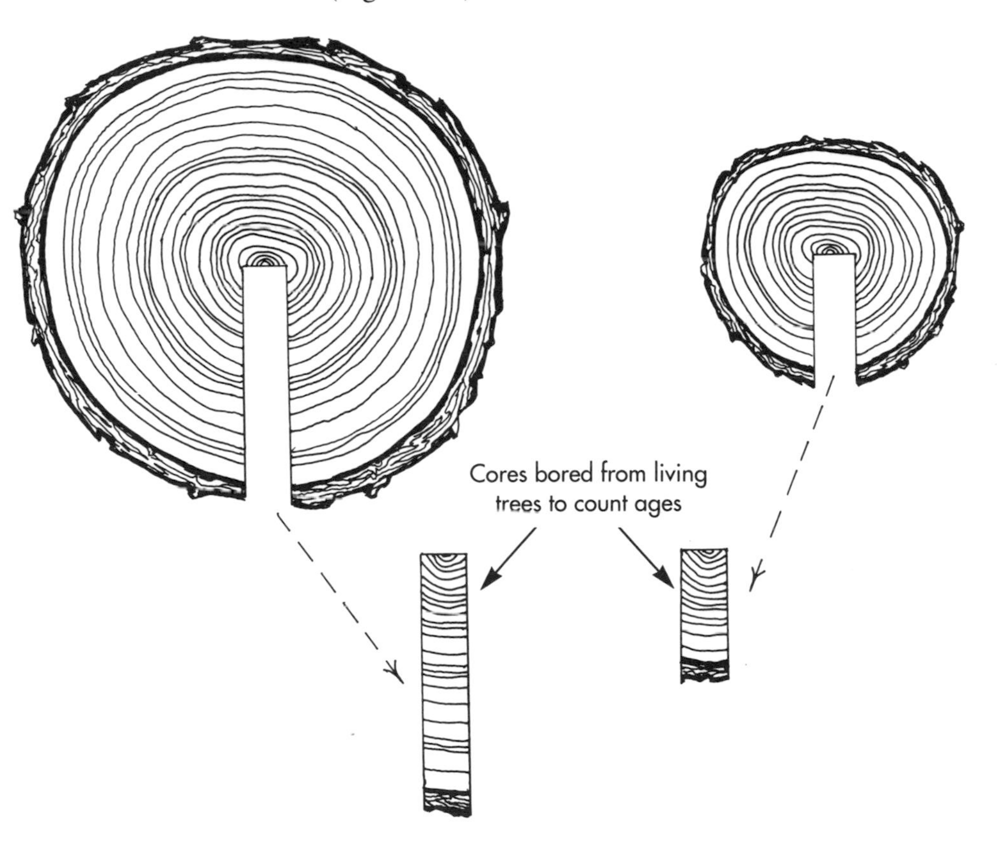

Figure 1-2. Because changes in width of the rings from year to year are produced by changes in growth conditions such as moisture and temperature, trees in the same locality usually have the same patterns of wide and narrow rings, as shown here. A tree need not be cut down to learn its age; instead, a "core" can be drilled out from which the rings can be counted.

Many kinds of trees live for several hundred years. Some large trees, such as the sequoias and redwoods of California, have been found to be several thousand years old. But such longevity is not limited to large trees. Counting growth rings has also shown the bristlecone pine, a small, scrubby tree that grows in the western United States, has been found to have astonishing ages. The tree in Figure 1-3 is about 1,600 years old. Another in the White Mountains of California was found to be 4,600 years old. A bristlecone cut down in Nevada in 1964 proved to be even more ancient, 4,900 years old. These gnarly old trees (Figure 1-3) were just beginning to grow from seedlings nearly 3,000 years before the birth of Christ. They were hundreds of years old when the ancient Egyptians first began to build their pyramids.

USING TREE RINGS TO DATE ANCIENT CIVILIZATIONS

Although the age of a tree is interesting to know, growth rings can be used to obtain much more important information. They can help scientists learn the dates of ancient civilizations. Our ancient ancestors used wood for beams in their buildings, and many samples of wood have been found in the ruins of settlements built thousands of years ago. Of course, counting the growth rings of these beams would tell us only the age of the tree when it was cut down. The number of rings will not tell us when the tree was cut down or when the building was constructed.

Climate Patterns Can Date a Log • It happens, however, that tree rings have other stories to tell. Trees do not grow the same amount each year; consequently, growth rings tend to have a different width each year. For example, in a year when the weather is warm and moist, growth is easy and the ring is wide. In a year when the weather is cold or dry, growth will be slower and a nar-

Figure 1-3. This bristlecone pine in Bryce Canyon, Utah, is only about twelve feet tall, but has a measured age of about 1,600 years.

rower ring will be formed. In other words, the pattern of narrow and wide growth rings in each tree shows poor and good years. Of course, just as weather patterns are irregular, so are the patterns of growth rings. For example, a tree ring pattern might show a poor year followed by four good years, then two poor years, ten good ones, and so on. However, trees from the same region (exposed to the same climate) tend to show the same pattern. By comparing the pattern of a tree whose cutting date is known with the patterns of older logs from the same region whose growing periods overlap with each other, we can obtain a continuous tree ring pattern that reaches back many years.

By combining that knowledge of regional tree ring patterns with careful detective work starting with modern logs, the dates for beams that were used in buildings back in prehistoric times have been found. Figure 1-4 shows a simplified example of how this process works.

Tree ring chronology, or dendrochronology, as this dating system is called, has helped to solve some mysteries about prehistoric Indian villages in the southwestern United States. Archaeologists excavating those ancient village sites had found artifacts such as pottery, cooking utensils, and tools. Such items told us much about what life was like in those days. But the people who lived in those villages—in fact, that entire early American civilization—left no written records. The unanswered questions were, when were those days?—when did these people live? And how rapidly did they progress from a hunting-gathering society to a farming, and later an urban, society?

To answer these and other questions, we needed a way of dating artifacts, villages, and happenings. One way of measuring time was found in wood samples from those settlements. Some were pieces of charcoal, others heavy beams that had been cut so long ago that they showed the marks of stone axes. Through careful detective work, the tree ring patterns were worked out as far back as 1,500

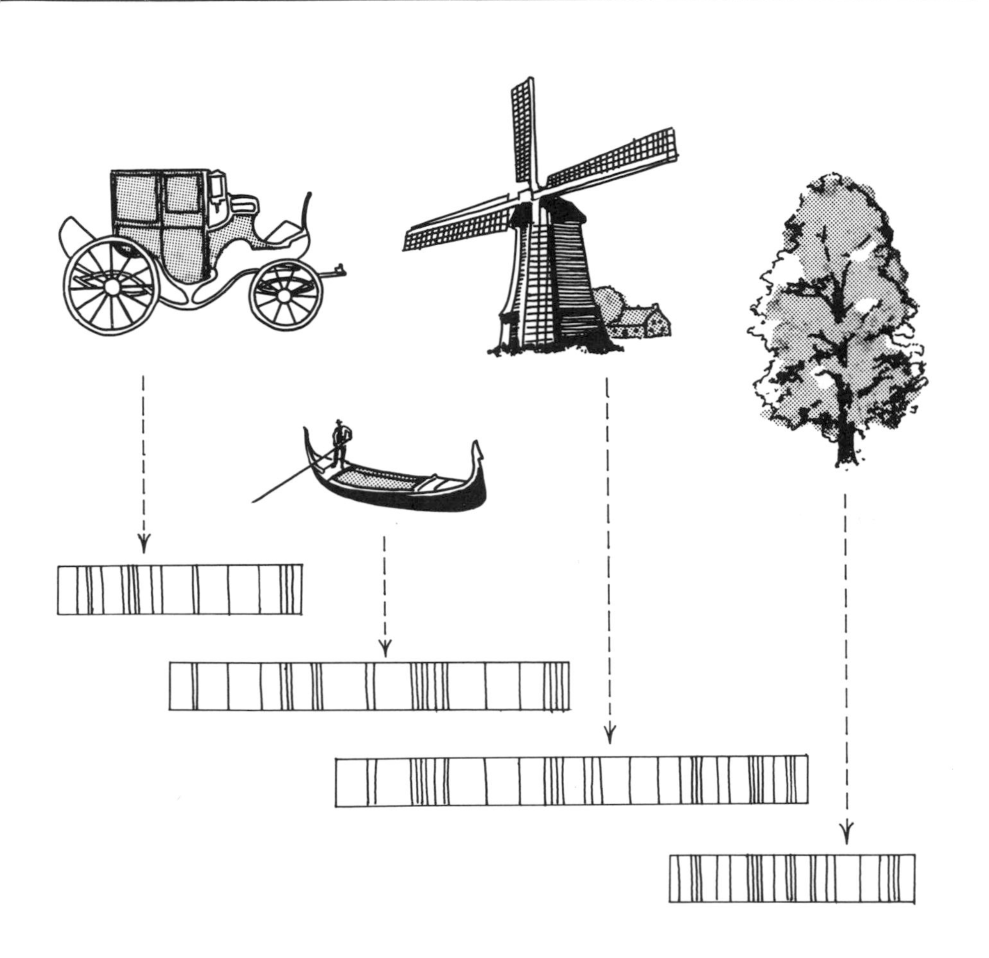

Figure 1-4. The use of year-by-year variations in growth rings from one log to another has made it possible to date beams of ancient buildings and other wooden artifacts.

years ago. We now know when these early Americans lived, and we have put together a cultural history of a civilization that left no written records (Figure 1-5).

Figures 1-5a, b, c, d. Wood beams like those above (the lower one cut to show clarity of rings) made it possible to date the ancient settlement at Mesa Verde, Colorado, (right) at around 1100 A.D.

1-5c

1-5d

Puzzles in Tree Rings • Despite their great value, tree rings have brought scientists enough puzzles to keep them busy and on their toes. For example, sometimes a ring for a particular year will be missing from a log. Sometimes two growth rings are produced during one year, when a tree stops growing for some reason and then starts up again. Sometimes a ring will not extend all the way around a tree. In such cases, counting rings in the wrong place on the cut log will give the wrong age. In cold climates, the growth rings of some trees are so close together that they are difficult to count. In the tropics, growth is more or less continuous all year round, so that clearly defined rings may be difficult to find. The loss of a year or two in age of the tree is unimportant, unless the gap makes it difficult or impossible to trace the ring pattern from one log to another.

An even greater problem is the question of whether the date of a timber is the same as the date of the erection of the building. Quite often, especially in areas where wood is scarce, beams were salvaged from old buildings and used again in new construction. The age of the beam, as determined by dendrochronology, might then be dozens or even hundreds of years greater than the building in which it is found. By comparing numerous beams and making careful checks, such incorrect dates can usually be detected and discarded.

OTHER INFORMATION FROM TREE RINGS

Tree rings are also used to study ancient climates. The narrow and wide rings from poor and good growing years are records of climate for the region (or perhaps for the whole earth) for those years. By examining tree ring patterns, climatologists can gather information on cycles and changes in climate that occurred over a thousand years ago.

For example, when the climatologist is examining tree

rings from a climate that he knows from other evidence was relatively dry year after year, he can assume that the change in width of the rings is telling him about changes in temperature of the region. That is, a series of wide rings signifies a stretch of warm years when growth was easy while a series of narrow rings signifies a period of cold years when growth was slower.

On the other hand, if the scientist knows from other evidence that the climate remained relatively constant year after year, he can assume that changes in the width of rings are telling him about changes in rainfall. That is, wide rings would indicate wet years, narrow rings dry years. Climatologists can thus find in tree rings records of climatic changes that go back far beyond any weather records made by humans.

Sunspot cycles have also been found to be recorded in tree rings. Sunspots, which are violent storms on the surface of the sun, increase in number to a maximum about every eleven years, then slowly decrease again. The increased radiation from these solar storms changes the amount of solar energy that reaches the surface of the earth. This in turn changes the climate of the earth enough to affect the growth of trees and other vegetation. From the growth changes recorded deep inside the trunks of trees, scientists can study and date the cycles of sunspots that occurred many hundreds of years ago.

DATING ARTIFACTS

Tree rings have not been very useful for obtaining dates for events that occurred more than about 1,500 years ago. The large beams needed to show tree ring patterns become more and more rare in the ruins of civilization farther and farther back in time. For one thing, wood rots away over many years unless it is preserved by unusual circumstances, such as an extremely dry climate. Also, the dwellings used by very ancient people were more flimsy and less permanent than more recent dwellings and had fewer heavy beams.

For dates beyond the limits of tree ring dating, we must turn again to the abundant supply of ancient objects called artifacts, found in the ruins of ancient villages, in graves, and in caves. Such artifacts, which may include stone tools, wooden implements, pieces of cloth, animal bone, pottery, charcoal, metal, and leather, have long told us how ancient civilizations developed. With new methods developed within the last thirty years, these items can now be made to tell us the dates of their manufacture.

These new methods came out of much earlier efforts to date the formation of the earth itself. We will therefore set aside the problems of dating bone artifacts and pottery for the moment and move backward in time in the next chapter, to the days when scientists first began to grapple with one of the greatest puzzles of all time: *how old is the earth and the life that inhabits it?*

2 THE AGE OF THE EARTH—THE OLD VIEW

Only a few hundred years ago, people believed that the earth had changed very little since the time when it was created. They knew that the mountains were made of rock; how could anything made of solid rock ever change? They knew that rivers eroded their channels as they carried sand and mud downstream, and that earthquakes sometimes shook things up a bit here and there. But they also knew that there was very little change in the landscape in a person's lifetime. Indeed, maps showed that the outline of the known world had not changed noticeably since the time of the ancient Greeks. And since the world had been created (they believed) only a few thousand years earlier than those ancient times, it must have been created essentially as they saw it then.

And yet—and yet when people looked at the earth carefully, they began to see evidence that the earth did not seem to be a clean, crisp, recent creation, but more like a pile of wreckage (Figure 2-1). Some observers began to see, too, that whatever process it was that churned the earth's surface into wreckage must have taken a very long time—much longer than a few thousand years.

THE PUZZLE: HOW OLD IS THE EARTH?

Is the earth a fresh, recent creation, or an ancient wreck? This was the question that a new group of scientists began to ask themselves less than 200 years ago. They began a new science called geology, whose purpose was to examine the earth to study its origin, its structure, and what had happened to its structure over the years. As the new geologists tried to understand the earth and its history, their thinking was hampered and confused by the belief, then held by nearly everyone, that the earth was not very old.

ANCIENT IDEAS OF THE AGE OF THE EARTH

Many people at that time had made guesses about the age of the earth. The Bible contained an account of the creation of the earth, but no date was given. The most famous estimate of the age of the earth was based upon the Bible. Bishop James Ussher (1581–1656), an Irish churchman and scholar, used the ages of the patriarchs whose stories are told in the Old Testament, starting with Adam, and other biblical events to calculate that the earth had been created in the year 4004 B.C., which is less than 6,000 years ago.

Other scholars came up with dates for creation as early as 7000 B.C. (9,000 years ago), perhaps arrived at by similar methods. Dates like these were widely accepted, and

Figures 2-1a, b. The rough, torn, and eroded surface of the earth makes it look more like an old pile of wreckage than a crisp recent creation.

were seriously quoted in books published as recently as a few decades ago.

Other ancient peoples had very different dates for the creation of the earth in their literature and legends. The Greeks thought the date was earlier than 9000 B.C., with some holding that the formation of the world was due to a natural (rather than supernatural) series of events. The Persians dated mankind's beginning at 12,000 years ago (from our time). The Egyptians counted 541 generations, about 10,000 years (before their time), to the beginning.

Peoples farther to the east believed the earth to be much older. The Chaldeans, who lived in the area of the Middle East now known as Iraq, believed the human race to be nearly half a million years old, and the earth 2 million. The sacred books of the Hindus declared the earth to be almost 2,000 million years old.

All of these guesses were, as we shall see later, far too low. The earth is now believed to be more than a million times older than Bishop Ussher calculated it to be.

SUDDEN CATASTROPHE—OR GREAT AGE?

As we saw earlier, the age of the earth then generally accepted—a few thousand years—caused the new science of geology a great deal of difficulty. As the new geologists examined the earth more and more closely, they found that the crust everywhere had indeed been eroded, churned, and twisted into a pile of wreckage. Because they accepted the current thinking about the earth's age, they had to conclude that all this must have taken place within a few thousand years. They therefore decided that the earth must have been wracked by some sort of sudden worldwide catastrophe. They presumed that this event took place after the "Great Flood" described in the Old Testament. After this more recent catastrophe, they said, the earth has remained the same to the present time.

Serious doubts about this catastrophe theory were raised by the growing evidence that the earth was very much older than was commonly believed. Some geologists attempted to explain away this evidence of age by declaring that the Creator had, for reasons of His own, created the earth with the "appearance" of great age.

GEOLOGISTS EXAMINE THE WORLD MORE CLOSELY

But the evidence that the earth is truly very old would not go away. Instead, it kept growing, with new facts popping up everywhere. Geologists began to understand the processes by which solid rock is washed away by rain and running water and redeposited somewhere else. They began to see that the present landscape of the earth was the result of these processes, and to realize that these processes are very, very slow. Geologists discovered that much of the crust of the earth is arranged like a many layered cake, with layers of various kinds of soil and rock stacked one on top of another. They found these layers in exposed cliffs, in banks cut by railroad and canal excavations, and in mine shafts.

THE DISCOVERY OF FOSSILS

In these layers of earth and rock, geologists made a most puzzling discovery: fossils. These bits of bone, shell, or stone seemed to be remains and imprints of animals and plants. Many of these remains seemed to be from life forms that were quite different from any now found on earth. But even more puzzling was their location. Shells from sea creatures were found many miles from the ocean, even high in the mountains. Bones and shells were also discovered buried deep in the earth, often imbedded in solid rock (Figure 2-2).

Figure 2-2a. This fossil (above) of a trilobite, a creature that lived at the bottom of ancient seas 250 to 550 million years ago, was found several hundred miles inland and more than 100 feet above sea level.

Early Explanations of Fossils • Could these objects really be the remains of ancient living things? If so, how did they get into such impossible places? All sorts of explanations were offered, many of them prompted more by belief than by scientific observation. Some investigators claimed that fossils had no life function, but had been created by nature "acting irrationally and unpurposively." A Swiss naturalist thought that fossil plants and animals had been placed in the earth by the Creator to display the harmony of His works "by a matching of the productions of land and sea." But why were some fossils broken and pressed into layers of rock? What about the fossil teeth, odd bones, and broken shells? How could they "display harmony"? Other investigators went so far as to suggest that fossils had been created by the devil to deceive, mislead, and perplex people.

Fossils were indeed misleading and perplexing, and drove some geologists back to the Scriptures for an explanation. In keeping with the idea that the earth was only a few thousand years old, their thoughts returned again to the Great Flood of Noah and they decided that fossils must be the remains of creatures killed in that disaster. They reasoned also that the Great Flood was one of the catastrophes that had caused violent changes to the crust of the earth. One investigator explained that the waters of the flood had dissolved stone and soil into small particles. This rock porridge was then mixed with the bodies (fossils) of

Figure 2-2b. This ancient herringlike fish (bottom, left) lived in the ocean over 50 million years ago. Its fossil was found in the Green River Formation high in the mountains of Utah.

drowned animals in a vast, churning mass. As the waters of the Great Flood subsided, the particles began to settle out, forming layers with the heavier materials on the bottom and the lighter materials toward the top.

Fossils and Layering • This process, if it really had happened that way, would have laid up the crust of the earth in smooth, regular layers like the layers of an onion. Questions immediately arose as to how this tale could explain the fact that many layers are bent, twisted, tilted vertically, or even turned upside down. This problem was explained away by postulating *another* great catastrophe, which had later shattered the carefully layered earth and churned it into a rough heap of ruins. In this catastrophe, the heaved-up strata became mountains, the downward buckles became deep valleys and seas, while the smaller crumples became hills and shallow valleys. Since the Great Flood and this later catastrophe, these investigators concluded, there have been no changes in the world.

Other geologists had already gathered enough information about the earth's crust to easily disprove this imaginative explanation. For one thing, water will not dissolve rock, nor will a "flood" easily break rock down into small particles. Also, the layers of the earth are *not* arranged with heavy materials below and lighter ones on top. And, most important, although fossils were being found scattered through many layers, no relics of any type from human civilization had ever been found in any of the very ancient debris that was claimed to have been laid down by the Great Flood.

The evidence was strong that what had happened on the earth could not have happened in one Great Flood or one great catastrophe of any kind. Further, the fossil record strongly suggested that most of the action—whatever it was—must have taken place long before there were people on the earth.

SLOW, STEADY GEOLOGICAL PROCESSES

Through such fact gathering and debate over many years, the science of geology slowly came to accept the idea that the earth was not a recent creation, but very, very old, and that great changes to its surface had occurred very slowly over a very long period of time. Water and wind had worn down the land and deposited the debris in low places in layers, new ones on top of older ones. These layers had often been pressed into new kinds of rock. Large areas of land had sunk beneath the sea and arisen again and again. Layers of rock had been bent, tilted, and even flipped upside down by the slow movement of the earth's crust (Figure 2-3). Although violent catastrophes such as earthquakes, floods, and volcanic eruptions had indeed occurred, their effects upon the earth's crust were local and relatively small.

Geologists recognized that the processes that had churned the crust of the earth were so slow that their effects could scarcely be detected over the lifetime of one person. It is almost impossible for us to tell, for example, whether a range of mountains is being pushed up faster than it is being eroded away, or vice versa. But these processes are going on today under our very eyes, although their effects will not be really noticeable for thousands of years.

These conclusions were first put into writing in 1785 by James Hutton (1726–1797), a British physician and naturalist, in what became known as the principle of uniformitarianism. This principle was later elevated to the status of scientific law by Sir Charles Lyell (1797–1875), a leading geologist of his time. This principle says simply that the development of the surface of the earth has been going on all through the ages without interruption, and that the process of very slow change that we observe today has been responsible for the present surface features of the earth.

The acceptance of the idea that the earth is very, very

old made it easier to accept the fact—then being learned through the study of fossils—that many creatures who once lived on the earth are now extinct. The idea that the earth was apparently much older than the human race also brought doubts about ancient beliefs that the earth had been created specifically for the immediate use of human beings.

GEOLOGISTS SHIFT FROM BELIEF TO OBSERVATION

As geologists began to accept these new views, the mention of scripture and scriptural dates slowly disappeared from their writings and symposiums. They went to work diligently on the problem of learning the actual age of the earth from the observations they were making. They had now discovered some of the "clocks" hidden in the rocks—such things as strata, fossils, and the processes of erosion and deposition—and had learned to read them well enough to be convinced that these clocks had been running for a very long time. Although the clocks had helped scientists to obtain much information about geological processes, they could not yet be read well enough to provide even a good guess as to the actual age of the earth.

More time would pass, and many other discoveries would be needed, before the true age of the earth would be learned.

Figure 2-3. These layers of sediments have turned to rock and have been fractured and bent.

3 LEARNING TO READ STRATA

When geologists began to look carefully at the layering of rocks and soil in the earth (called strata), they found that such strata exist nearly everywhere. Cliffs, riverbanks, quarries, mines, and mountainsides all show layers of rock and soil, one on top of another. Today you can see glimpses of the earth's strata wherever highways and railways cut through hillsides or mountains (Figure 3-1).

WHY IS THE EARTH'S SURFACE DIVIDED INTO LAYERS?

What does this strata mean? That was the question to be answered by the scientists who prowled the earth seeking to understand the layering. The idea that the strata might be sediments that had been carried by running water and laid down in orderly fashion, with the oldest on the bottom and the newest on top, was proposed by several people as long ago as the 1700s. But it was William Smith (1769–1839), a British civil engineer and surveyor who worked on both mines and canals, who first awakened the new

Figure 3-1. The light and dark areas in this photograph were sedimentary layers that have turned to rock. The strata have also been tilted and fractured.

science of geology to this principle—the principle of superposition. The strata lay, he said, "like slices of bread and butter" in a definite, unvarying sequence in a given region, with the younger layers resting on top of the older layers. Although the principle of superposition seems simple and obvious to us now, it was a very important discovery at that time.

There were some strata, though, that didn't seem to follow that principle. "Cuts" that you can see from your car window along almost any highway in hilly country will show strata tilted at odd angles, sometimes tilted vertically. Which way is up? Geologists have been fooled by strata that have been turned completely upside down; that is, upper layers have become the lower layers, and vice versa.

WHAT ARE FOSSILS, AND WHAT DO THEY TELL US?

Fortunately, most strata contain fossils, which, as we learned in Chapter 2, were at first a great puzzle in themselves. As soon as fossils were proved to be the remains and signs of ancient life, they were used to help solve the riddles of the strata.

William Smith was also one of the first geologists to study fossils closely and objectively. When he examined fossils collected from various strata, he soon realized that each layer usually contained different kinds of fossils. He had stumbled upon the important fact that the different plants and animals whose fossils he found had followed each other in the world's history in a definite, recognizable order. Smith and other geologists discovered that fossils found in the upper strata looked most like modern animals, while the fossils from the deeper layers were different and strange.

Geologists could not at first explain why different fossils were found in different strata. The idea was still strong in people's minds that living things, like the earth itself, had been created recently and all at the same time.

On the other hand, the newly discovered fossils seemed to show that living things had somehow *changed* as time passed. Some creatures of the past seemed to have disappeared from the earth entirely!

The Problem of Belief • Many people were still trying to adapt what they *observed* to fit what they *believed.* As a result, some strange and imaginative ideas were proposed to explain why different fossils were found in different strata. One geologist suggested that there had been a series of "revolutions" which had destroyed all the animal life in a particular area. The area was then "taken over" by other (slightly different) animals which moved in from nearby regions.

Another investigator suggested that there had been not just one "creation" but a whole series of them since the beginning. From his study of fossils, he identified some twenty-seven different times that animals must have become extinct and were then replaced by the creation of new ones. Just how or why all animal life had been wiped out and re-created in slightly different forms, he didn't say.

Darwin and Evolution • This sort of speculation became unnecessary after Charles R. Darwin (1809–1882), a British naturalist, put forth his theory of evolution and natural selection. This theory, which was based upon data gathered over many years of observation, not only offered evidence that living things change but also described the process of natural selection that controls these changes. Darwin's theory, which has long been accepted by science, thus explained the difference in fossils that occurs between successive strata to be a record of the slow changes that have occurred in living things throughout earth's history. The fossils in old (lower) strata are unfamiliar because they are early forms of life, now extinct. The fossils in newer (upper) strata tend to be more familiar because they are near-ancestors or close relatives of present creatures.

The New Science of Fossils • In their investigation of fossils, geologists began a whole new science, now called paleontology, which studies fossils and the ancient life they represent. Fossils of many species could often be arranged in order of the changes that occurred during their time on earth, from earliest appearance to time of extinction, or to present-day forms.

Fossils and Order of Strata • As the fossils found in various strata were sorted and arranged according to age, geologists found that they had an excellent tool for settling questions of the sequence of strata, that is, which stratum was older, if doubt existed. But even more important, they had a tool for recognizing strata that were of the same age, even though found in areas that were far apart. If two strata contained fossils of the same creatures, it could be assumed that the two were laid down at about the same time in the earth's history, and might in fact be parts of the same widespread layer. (The process of determining the relationship of one stratum to another—for example, which came first—by using the presence of fossils, or by other methods, is called correlation.) Fossils and the organic evolution they illustrated thus gave valuable information on the order in which events happened, but no dates.

As scientists studied the evolution of creatures from the fossil records, they found that the rate of evolution was not constant. Evolutional changes to some forms of life seemed sometimes to proceed at an explosive rate, then to stop for long periods of time. Why this is true remains largely a mystery to this day.

Some fossils proved to be more useful than others for tracing strata from one point to another on the earth, and for determining which stratum is the older. Organisms that changed very slowly over time are less useful for dating because their fossils may all look the same in several strata laid down over very long periods. But organisms that evolved (changed) very rapidly left behind fossils that are clearly different in successive strata. These differences make it easier to determine the order of strata and to identify the same layers in different areas. If the creatures had also spread rapidly over a very wide area of the earth as they evolved, their fossils were doubly useful: they could also help geologists to identify—or correlate—the same layers over that whole area.

With these new understandings of strata and fossils, geologists found that they could correlate geologic formations over vast areas of the earth's crust. They could tell, in other words, that an outcropping found in one place had been laid down in the same time period as one discovered hundreds or even thousands of miles away. Such information made it possible to prepare geological maps that explained and illustrated the long, slow processes that had churned large areas of the earth's surface to produce the landscape we now see.

DIVIDING EARTH'S HISTORY INTO GEOLOGIC PERIODS

By fitting together the succession of all known strata, the kind of rock in the strata, and the fossils found in each, geologists gradually classified the strata of the earth in order of their age. The various periods of time in which each stratum was laid down were given names taken from the names of the places where these rock formations were first discovered or are particularly well displayed. These geologic periods were then arranged in order and grouped into eras whose exotic names have simple meanings: Paleozoic ("ancient life"), Mesozoic ("middle life"), and Cenozoic ("recent life"). The Precambrian era is that period from before the appearance of life (fossils) back to the formation of the earth's crust. It is the longest era by far—approximately seven-eighths of the earth's history.

The "Geologic Column" • Figure 3-2 is a chart showing the (standard) geologic column. This "column" was an important development, because it divided earth's

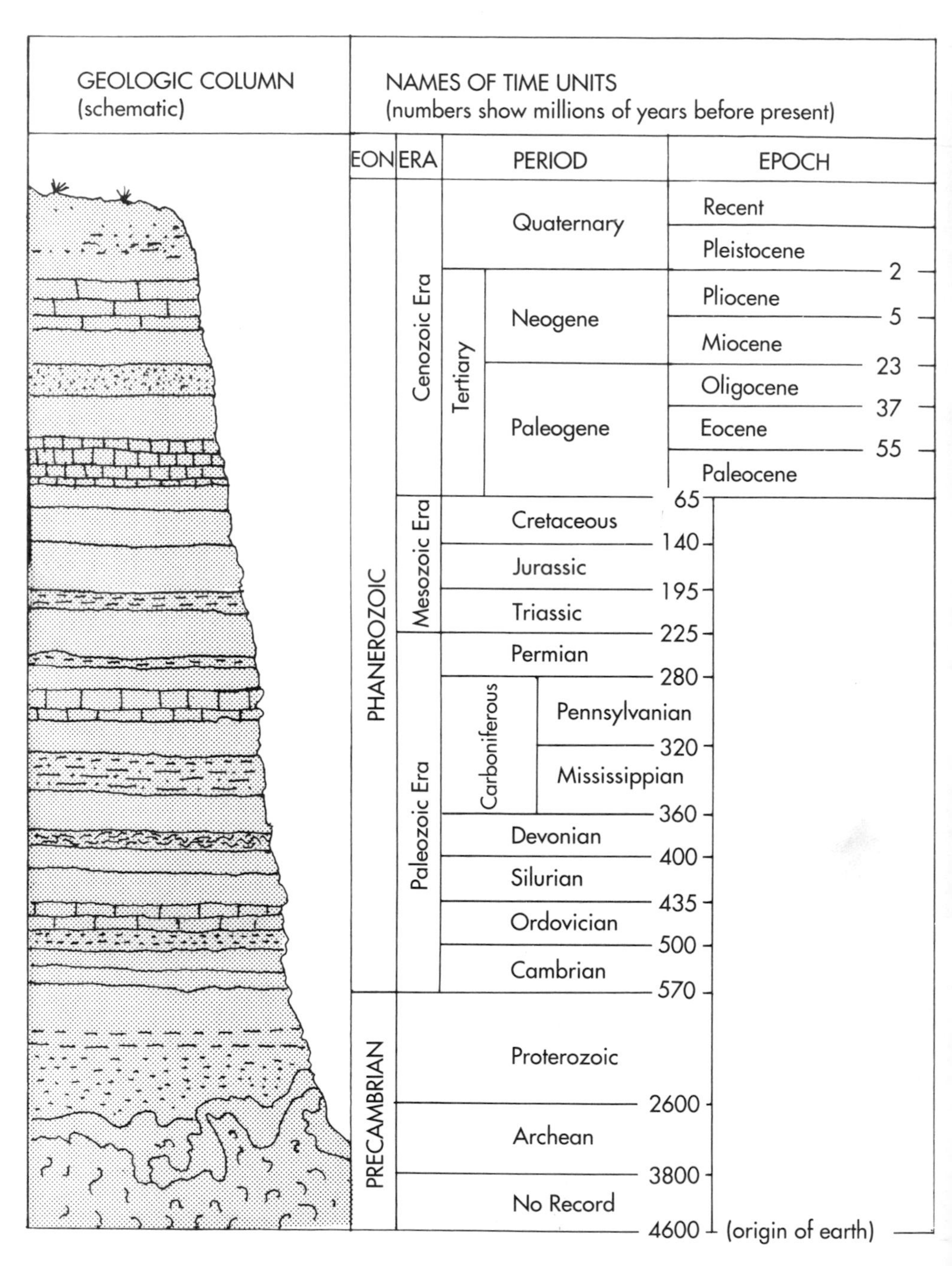

Figure 3-2. The standard geologic column is an actual stack of deposits (shown schematically at left), grouped into eons, eras, periods, and epochs.

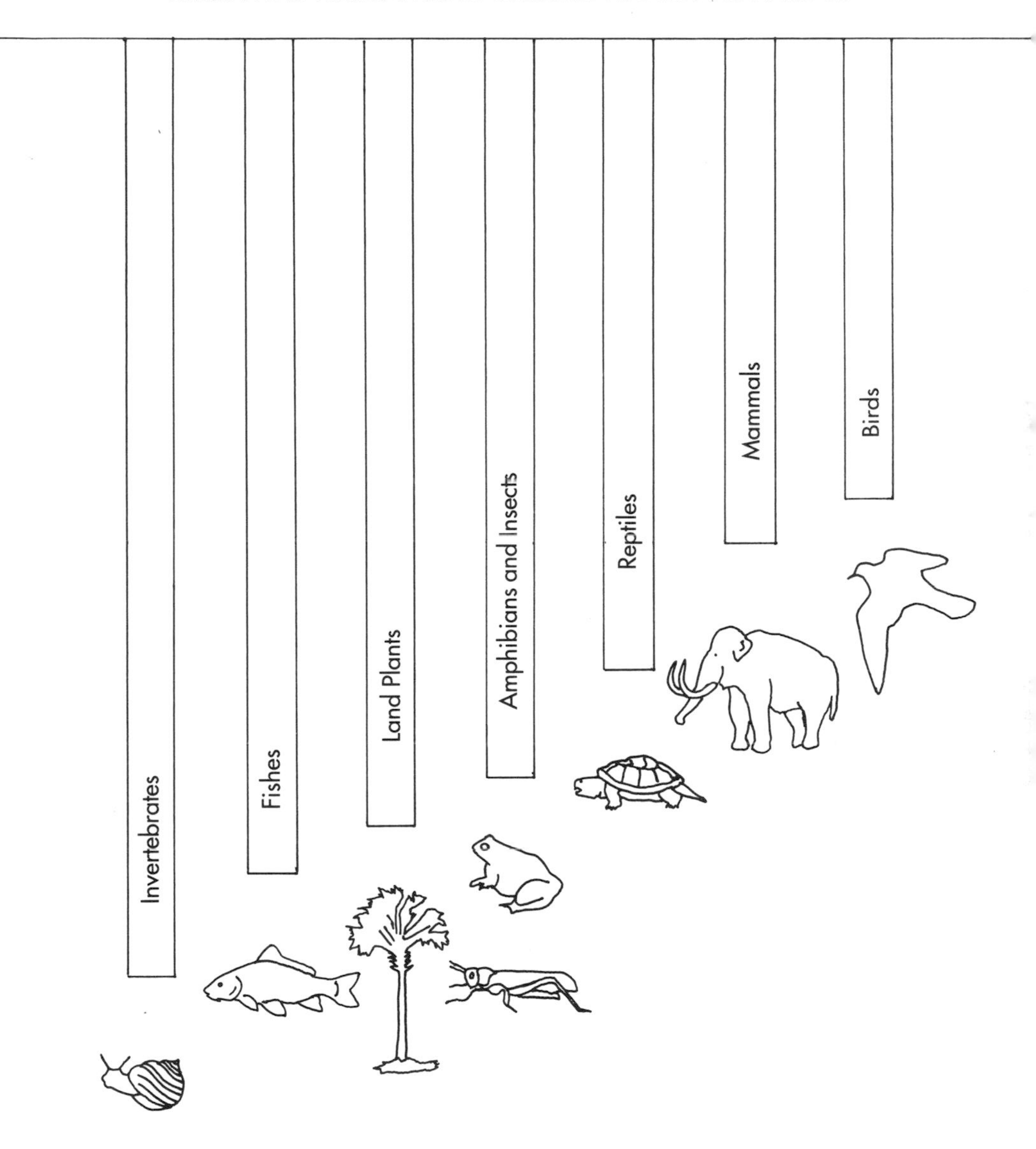

The origin of various forms of plants and animals are shown.

history into convenient periods for detailed study. It had one serious flaw at that time—there were no dates for the periods it listed. No one knew—or could even make a good guess—when each era had occurred or how many years it had contained.

THE NEED FOR DATING THE GEOLOGIC ERAS

Scientists had learned a great deal about what had happened to the earth's crust, and the order of these events, but next to nothing about *when* these things had happened. It was clear that some method of dating was urgently needed if we were to know all that we wished to know about the history of the earth and the development of living things. And so the attention of scientists—not only geologists but physicists, mathematicians, and naturalists as well—turned anew to the old problem of investigating that old, stubborn, unsolved mystery: the age of the earth.

Using Erosion and Deposition as a Clock • Fairly early in the history of geology, scientists tried to use geological events themselves as "clocks" for measuring geologic time. For example, they tried to find some idea of the age of the earth by estimating the time needed for the earth's crust to be eroded away and deposited as new strata. The rates at which the wearing-away and the depositing occur are so slow that they are very difficult to measure or even to estimate. These processes also vary greatly from one place to another, depending upon such things as how hard the rock is and how much water is flowing over it.

The rates at which sediments are deposited can also be very different from one situation to another. The rates can vary from a fraction of an inch per century to many feet per hour! And the process may sometimes stop completely for periods even longer than the periods of deposition. Because of these problems, no good estimates of the earth's age came from efforts to use erosion and deposition as a dating clock, beyond a confirmation of the idea that the earth is probably very old.

In order for a geologic (or any) process to serve as a useful clock, three things are necessary:

1. The starting condition must be known
2. The rate at which the change occurs must be known, and preferably be a constant
3. The final condition must be known

The Candle as a Clock • Before mechanical clocks were invented, candles were sometimes used for measuring the time of day. As the list below shows, the conditions necessary for measuring time from a burning candle (Figure 3-3) are easily established and correspond to the requirements listed above:

1. The length of the candle at the start can be measured
2. The rate at which the candle burns is constant and can be measured (against an hourglass or the movement of the sun, for example) in advance
3. The final length of the candle can be measured

A candle can thus be marked with "hour lines" or some other measure from which elapsed time can be observed directly as the candle burns.

If we apply these conditions to the problem of measuring time by measuring erosion and deposition, we can see that such a process might be somewhat useful for studying a single formation, such as the Grand Canyon. We can see what the canyon looks like now and measure its present depth accurately (item 3 above, the canyon's final condition, for our purposes). We may also have a good idea what the area looked like when the erosion process started (item 1 above, the canyon's starting condition). But we don't know accurately the rate of erosion (item 2 above); for example, how many inches are worn

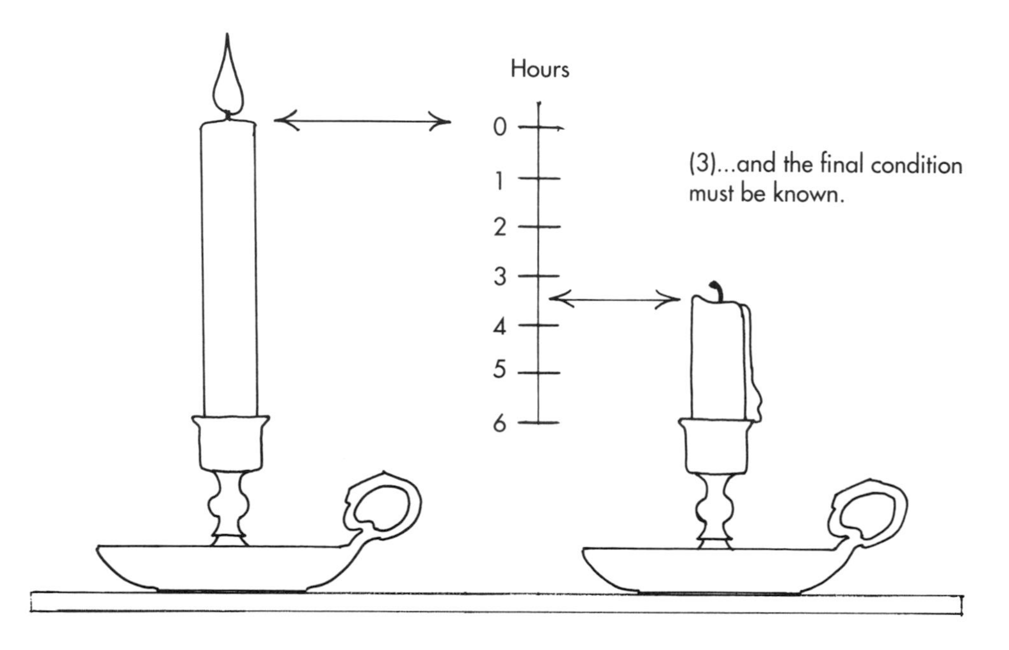

Figure 3-3. Before clocks were available, candles were sometimes used to measure elapsed time.

away per thousand years, although we might make an estimate from a study of water flow, the hardness of the various layers of rock, etc.

Such a method might thus yield a rough estimate of the time required for a particular geologic formation, such as the Grand Canyon, to be dug out by erosion. But it seems clear that estimating the amount of time required to erode, redeposit, and erode, over and over again, the entire surface of the earth to a depth of many miles would be an impossible task.

Clocking the Movement of a Glacier Long Ago • One of the earliest attempts at using deposition as a clock did produce useful results for a very special period of geologic history: the ice ages. This clever method, called "varve analysis," actually made it possible to learn how many years it took for certain geologic events to take place. It was developed by Baron Gerard de Greer in Stockholm, Sweden. He discovered in the 1870s that there was a regularity in the layers, or varves, found in certain claylike deposits laid down by melting glaciers. (The word *varve* comes from the Swedish *varve,* "a layer.") Close study showed that these varves could be counted, compared, and coordinated over an area to give both a time scale and motion scale to the melting of glaciers.

A pile of rubble, called a moraine, which has been pushed along by the forward edge of the advancing glacier, is dropped at the point of farthest advance of the glacier. This moraine acts as a dam to hold the meltwater lake that grows as the glacier melts and recedes. As the glacier melts to positions 1, 2, 3, etc., year by year, the fine sediments released into the water by the melting ice settle to the bottom. In summer, when the ice melts rapidly, a large amount of sediments is released. The coarser grains fall quickly to the bottom, and are covered later by the finer sediments which settle out much more slowly. Then the process stops when winter comes and the lake freezes. When melting resumes in the spring, the sedimentation process starts again and deposits another layer of coarse, then fine, sediments.

The different layers are generally distinguished by color, and can be counted and measured for thickness. The thickness of varves varies from one year to the next, with thick varves tending to occur during warm, long summers when the melting rate is high. Thin varves occur during cool, short summers. This annual layering, and the pattern of variations that occur from year to year, are thus analogous to the patterns found in tree rings. The varve patterns can be used to match up (correlate) varve measurements from different locations, which makes it possible to locate the edge of the retreating glacier over a wide area.

By counting varves and measuring the travel of the end of the varves across the meltwater pool, the rate at which a glacier has melted and retreated can be measured with good accuracy. Unfortunately, lakes and varves are not found for all parts of a glacier's retreat; blank spaces in the varve measurements must be filled in with estimates.

By correlating varves from one pool or lake to the next, and filling in with estimates where no record can be found, the retreat of the ice sheet from Long Island through New England to James Bay, Canada, and on to its present location in northern Canada was estimated to have taken over 36,000 years. Other researchers have reduced this number somewhat, as the search for a time scale for this "recent" geologic event goes on.

One of the most difficult tasks in varve analysis is hooking up the varve period with the present time, or to some other known date. However, as in the case of strata dating, the scientist is happy to have some numbers, however tentative, in a place where he might otherwise have no numbers at all. And there is always the possibility that a "missing link" will be found that will tie together a whole chain of dates.

Using the Earth's Heat Loss as a Clock • For the central task of learning the earth's age, there were several other geologic processes that seemed to have good possibilities. One of these was suggested by the fact that the temperature of the earth has been measured to be greater at deeper and deeper levels below the surface. William Thomson (Lord Kelvin, 1824–1907), a British mathematician and physicist who was very much interested in the science of heat, suggested that this temperature gradient could be used as a clock to measure the age of the earth.

The earth was once a molten ball, he believed, that solidified long ago and then cooled down to its present temperature. It is still cooling, he reasoned, since the higher temperatures measured in deep mines showed that heat was still working its way to the surface. Starting with (1) from the list on page 43 and on Figure 3-1, the earth at the temperature of molten rock, and (2) an estimate of the rate at which the earth's heat would leak away into cold space, Thomson calculated the time required for the earth to cool to its present temperature (3).

From these calculations, he concluded that the earth must have been formed not less than 20 million years ago, and probably not more than 400 million years ago. His assumptions and calculations clearly contained much uncertainty. Although the answer he got did not even approach the actual age of the earth, it did bring the world to thinking about the earth's age in terms of many millions of years. We now know that his numbers are far too low, and we know what spoiled his calculations. He assumed that the heat still inside the earth is only that left from its former molten state. We now know that assumption was incorrect because he was not aware of the heat that radioactivity has generated inside the earth all through its history, and is doing so today. (Kelvin recognized this possibility, good scientist that he was, when he stated in his writings that his calculations might be invalid if "sources of heat now unknown to us" were involved.)

Geologists now believe that the earth was once very hot but probably not completely molten, and actually lost that heat a long time ago. The heat still coming up from the interior, they say, is largely from the energy given off in the decay of radioactive elements such as uranium. Thomson's calculations were thrown off just as a calculation of time from the candle in Figure 3-3 would be if an unknown amount of wax were somehow added to the candle as it burned.

Using Ocean Salt as a Clock • In the early 1900s several scientists tried to use the process by which the oceans became salty as a geologic clock to measure the age of the earth. If it is assumed that the oceans were originally fresh water, and if the rate at which salt is dissolved from the land and carried to the ocean can be estimated, they reasoned, the time that it took for the oceans to acquire their present salt content could be calculated. This time would be the age of the earth.

Unfortunately, it proved difficult to make a reliable estimate of how rapidly salt is brought to the ocean. Also, it was not known then that not all of the salt washed out of the land has remained in the ocean. Some of it, we know now, has been redeposited in thick salt beds and the "salt domes" that are well known in the world's oil fields. These salt layers, now deep in the earth, were deposited when ancient seas dried up and the area was covered by new sediments. These conditions make both items 2 and 3 in Figure 3-3 uncertain.

All calculations of the age of the earth made by this method were very low—one was only 80 million years. More recently, new calculations based upon ocean salt have been made that included corrections for some of the uncertainties in earlier work. These calculations gave a minimum age of 700 to 800 million years, with a possible maximum of 2,350 million years. These values are closer to the true age, but still contain a large measure of uncertainty.

Attempts to learn the age of the earth by studying erosion, heat loss, ocean salt accumulation, and other processes have thus given unreliable answers. Although most of them greatly underestimated the age of the earth, they did help to prove that the earth is not a recent creation, but is actually very old, with an age measured in hundreds or even thousands of millions of years. In the early 1900s the generally accepted estimate for the age of the earth had risen to 100 million years, although many scientists by that time suspected that the earth might be much older.

THE DISCOVERY OF A RELIABLE CLOCK FOR DATING

A few years before the turn of the century, a new discovery was made in physics that would turn the world of science upside down. This was the discovery of radioactivity by Antoine Henri Becquerel (1852–1908), a French physicist. Within half a dozen years, a British physicist, Ernest (Lord) Rutherford (1871–1937), applied the new discovery to calculate the age of a sample of rock containing uranium. The answer he got was 700 million years—seven times the age of the earth accepted at that time! The new science of radioactivity not only upset all earlier ideas of the earth's age but also seemed to promise what geologists had been looking for: an accurate means for dating events that had occurred millions and billions of years ago.

CLOCKS IN EARTH ROCKS

THE INDESTRUCTIBLE ATOM

At the time of Becquerel's discovery of radioactivity, scientists knew that all matter was made up of atoms and they had some knowledge of what atoms were like. But until that time, they had believed the atom to be eternal, unchanging, and indestructible. They knew that atoms of one element could be combined chemically with atoms of another element to form a chemical compound. Atoms of hydrogen, for example, could be combined with atoms of oxygen to form molecules of the compound water. Atoms of uranium could be combined with atoms of oxygen to form uranium oxide. But no atom of any element could be changed into an atom of some other element, so far as anyone then knew.

THE DISCOVERY OF RADIOACTIVITY

Yet such a change, Lord Rutherford explained a few years later, was behind Becquerel's discovery. Radioactivity, he declared, is the result of the spontaneous disintegration of an atom of one element into an atom of a wholly different element.

The hope that one element could be changed—or "transmuted"—into another element was behind the work of the ancient alchemists, who had labored fruitlessly for centuries, trying to change a base metal, such as lead, into a precious one, such as gold. Because science had come to the conclusion many years earlier that such transmutation was impossible, scientists greeted Rutherford's explanation of radioactivity with doubt and opposition rather than approval. After having been convinced that the best efforts of chemistry and physics could never change lead into gold, scientists were now being asked to believe that some elements could spontaneously change from one kind to another, without any help at all! It was too much to expect.

But Rutherford was correct, and within a few years experiments by other scientists showed that the element uranium did indeed change, through a series of "radioactive decays," into the element lead. Rutherford was also correct in predicting that radioactive decay might provide the kind of "clock" that geology so badly needed to date the rocks of the earth. The changing of uranium into lead seemed to be a steady, reliable process, far more reliable than the erosion of the land, the cooling of the earth, the accumulation of salt in the ocean, or any of the other "clocks" that geologists had tried to use.

HOW RADIOACTIVE DATING WORKS

The idea behind radioactive dating is simple enough, though a little more complicated than the process described in the candle analogy in Figure 3-3 (Chapter 3). In the case of uranium, the rate at which it decays into lead has been measured and is known. If the amount of both lead and uranium in a rock can be measured, then the number of years required for the lead to be produced by the decay of uranium can be calculated (Figure 4-1).

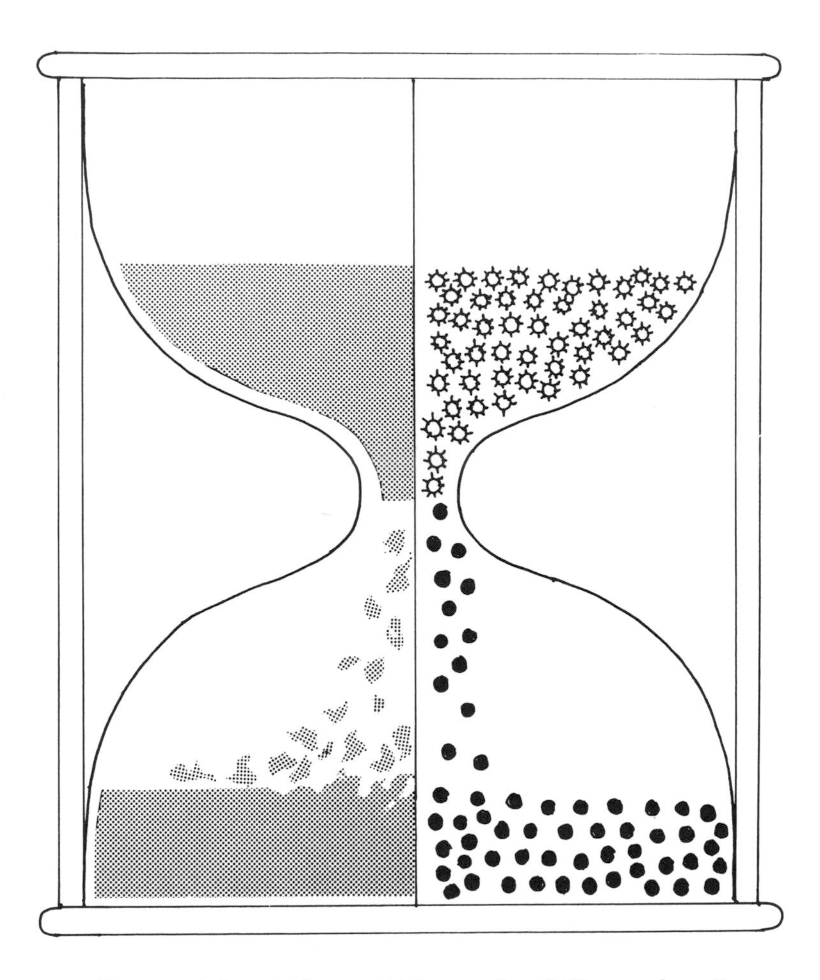

Figure 4-1. (Above) Schematic of "hourglass" analogy showing measurement of short periods of time by amount of sand flowing, and long periods of time by measuring amount of uranium 238 that has decayed to its daughter product, lead.

To understand fully the process by which an atom of an element changes when it "decays" into an atom of a different element, an understanding of the makeup of the atom and how its component parts are involved in chemical and nuclear processes will be useful. Readers who wish to review this background information will find it in the Appendix.

Figure 4-2 lists each step of the entire decay process for uranium 238, which ends in the production of the element lead 206, which is stable and will decay no more.

ISOTOPE (symbol)	ELEMENT (Name)	DECAY PROCESS*	DECAY PRODUCT (symbol)
^{238}U	Uranium	alpha	^{234}Th
^{234}Th	Thorium	beta	^{234}Pa
^{234}Pa	Protactinium	beta	^{234}U
^{234}U	Uranium	alpha	^{230}Th
^{230}Th	Thorium	alpha	^{226}Ra
^{226}Ra	Radium	alpha	^{222}Rn
^{222}Rn	Radon	alpha	^{218}Po
^{218}Po	Polonium	alpha	^{214}Pb
^{214}Pb	Lead	beta	^{214}Bi
^{214}Bi	Bismuth	beta	^{214}Po
^{214}Po	Polonium	alpha	^{210}Pb
^{210}Pb	Lead	beta	^{210}Bi
^{210}Bi	Bismuth	beta	^{210}Po
^{210}Po	Polonium	alpha	**^{206}Pb** (Lead)

*Alpha decay takes place through ejection of a proton from the nucleus; beta decay, through ejection of an electron (positive or negative) from the nucleus.

Figure 4-2. (Right) Uranium 238 decays through all of the radioactive elements shown on this table to become lead 206. For each step, read across. The type of radiation for each decay step is shown in the third column. The decay times for some steps are very short; for others, very long.

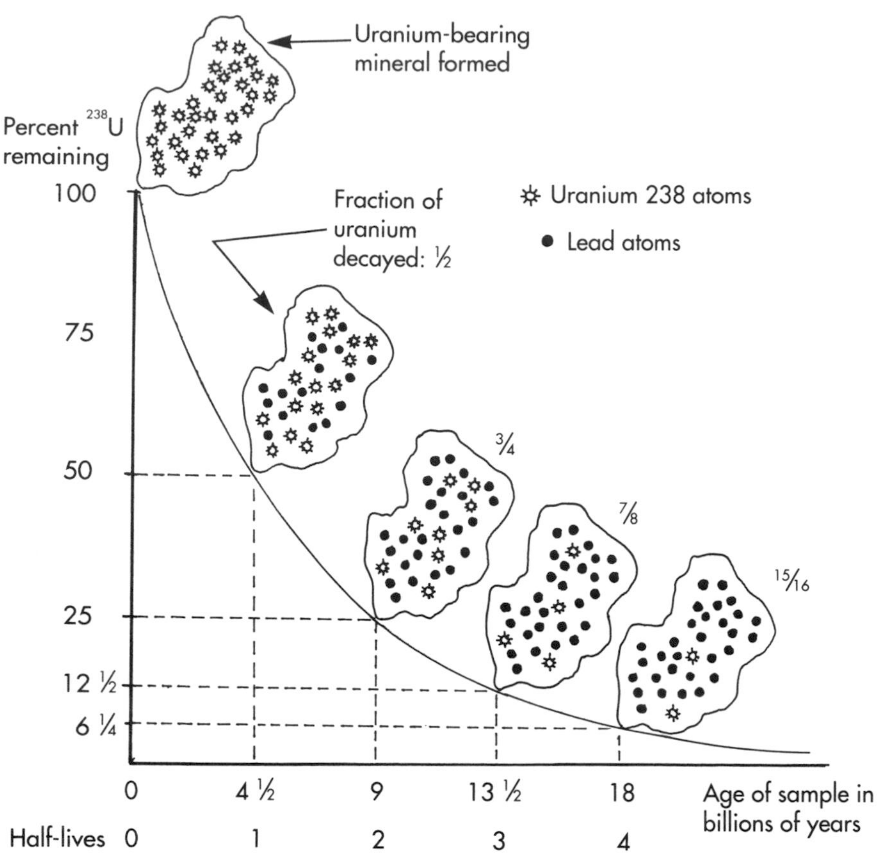

Figure 4-3. Schematic diagram of decay of a crystal containing U 238, showing half-life concept. Example: If half of the uranium in a sample is measured as having decayed, enter the chart at 50 percent on the vertical scale, then move across to the curve and down to read on the lower scale an age of four and a half billion years.

The length of time required for one-half of any group of uranium atoms to decay into lead is called the half-life of the element (Figure 4-3). The half life of uranium 238 is about 4½ billion years; that is, in any sample of uranium, half of the uranium atoms will decay to lead in 4½ billion years, and half of the remainder will decay in the next 4½ billion years, and so on. For dating long geologic periods, we are interested in elements that have a long half-life, such as uranium.

OTHER RADIOACTIVE ELEMENTS ARE USED

Uranium is perhaps the best-known radioactive element because of its use in nuclear power plants. But it is only one of about sixty-five radioactive elements or isotopes that are found in the rocks of the earth. (Atoms of the same chemical element that have the same number of protons in the nucleus but different numbers of neutrons are called isotopes. Isotopes of a single element differ in mass but have identical chemical properties.) Elements that are often used in radioactive dating are shown, with their half-lives and ''daughter product'' (the element, or isotope, produced by decay of a radioactive element), in the table below.

Element	Half-life in Years	Daughter Product
uranium 238	4,500 million	lead 206
uranium 235	713 million	lead 207
thorium 232	14,000 million	lead 208
potassium 40	1,300 million	argon 40
rubidium 87	47,000 million	strontium 87

Although uranium and other radioactive elements are not really plentiful, they do occur in very small quantities in most of the rocks of the earth. Measuring minute quantities of radioactive elements and their daughter products in any sample of rock is difficult at best. It is made especially difficult when several different isotopes of the elements, such as those shown for uranium and lead in the table, are present. Different isotopes of an element, as explained in the Appendix, have identical chemical characteristics and cannot therefore be separated or distinguished from each other by ordinary means. Fortunately, isotopes of an element do have one important difference: their atoms have different masses (or weights) because they contain a different number of neutrons. What was needed to sort one isotope from another was an instrument that could distinguish one atom from another on the basis of their masses.

THE MASS SPECTROMETER TO THE RESCUE

This sounds like an impossible task, and so it was, until the invention of the mass spectrometer. Just before the turn of the century, Wilhelm Wien, a German physicist and Nobel Prize winner, discovered that beams of charged particles could be deflected by a magnetic field. A decade later, J. J. Thompson, the British physicist who discovered the electron, used this principle to develop the first mass spectrometer. Over the next fifty years, this instrument was developed to the point where it is now a vital research tool in many fields of science.

The principle of the mass spectrometer is basically simple. A beam of ions (electrically charged atoms) from the material to be analyzed is accelerated through a vacuum chamber by an electric field. (An ion is an atom or molecule that has gained or lost electrons and therefore possesses an electric charge.) When this stream of ions is propelled through a magnetic field, the field exerts a force on them and bends the path of the atoms. As we might expect, the less massive atoms are deflected more easily while the more massive atoms, having more inertia, tend to resist deflection and continue on a path closer to the original straight line (Figure 4-4). Hence, when the ions strike a screen (or a photographic plate), the beam has been "sorted" so that the ions of different mass strike in different places.

From the pattern on the screen, and other information, the researcher can find out what kinds of atoms and how much of each are in the sample. In about 1939, the mass spectrometer was first used to measure the amounts of the various isotopes of uranium and lead in mineral

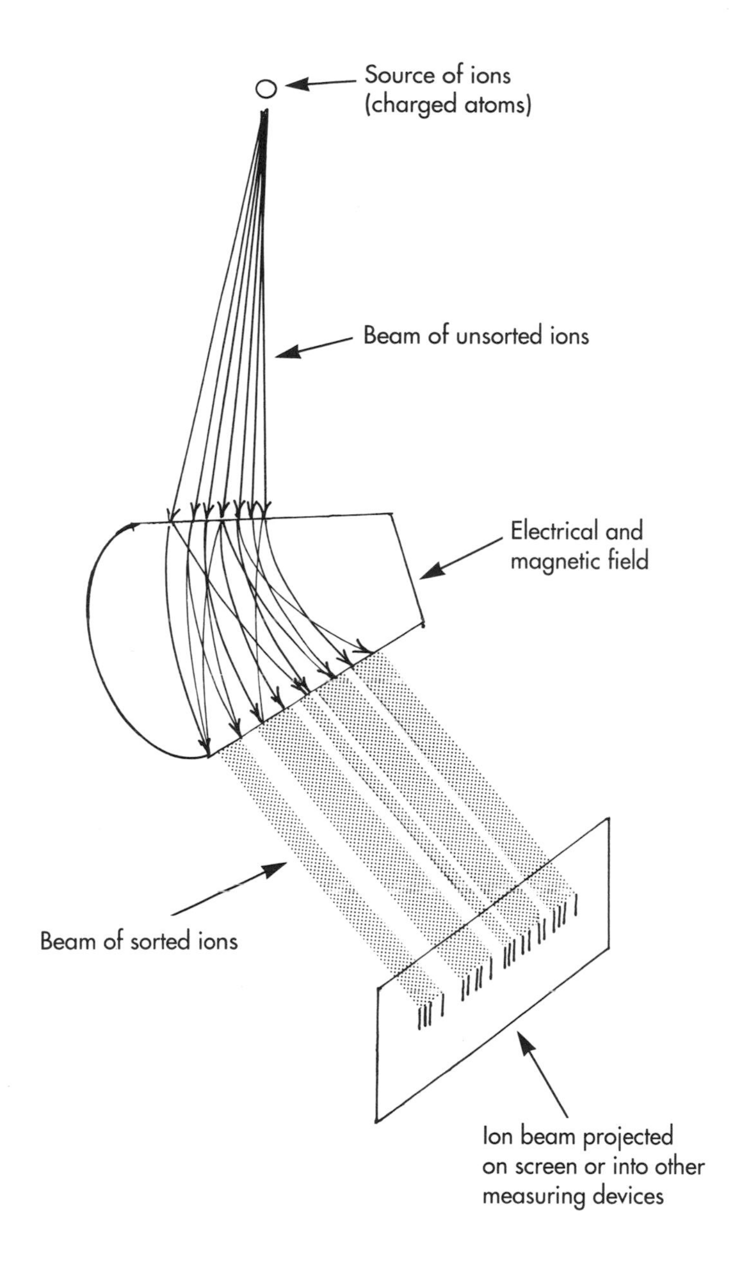

Figure 4-4. Schematic diagram showing principle of mass spectrometer and mass spectrograph

samples. These measurements showed that rocks could be accurately dated by radioactive-dating methods.

THE AGE OF EARTH ROCKS INCREASES *AGAIN!*

Geologists quickly put the new dating method to work on the problem of dating the earth's rocks. One experimenter collected samples of uranium ore from all over the world and analyzed them for uranium and lead content. When he calculated the amount of time it had taken for the lead to accumulate from the decay of the uranium, the answers he got were all more than *2 billion* years!

This answer was yet another astonishing increase in the age of the earth, and it was not readily accepted. First believed to be only a few thousand years, the age of the earth had climbed slowly to millions of years, stopping for a time at about 100 million. Then suddenly the new science of radiometric dating of some rock samples had suggested the age to be as much as 700 million. Now it was being increased to 2,000 million, or 2 billion, years. Even scientists sometimes resist information or sudden changes in ideas that they don't expect.

The new ages measured for the minerals of the earth caused such chaos in the scientific community that a committee was appointed around 1923 to look into the new method and see what it was all about. This action stimulated research, communication, and cooperation among scientists, and it was soon clear to nearly everyone that the new ages measured for earth rocks were undoubtedly correct!

This discovery dropped a new question on the world like a clap of thunder: if some rocks are over 2 billion years old, then how old is the earth itself? The only answer at this time, of course, was that the earth must be even older than the oldest rocks—perhaps a great deal older than 2 billion years. Finding out just how much older led to a vigorous search for the oldest rock available, as we shall see later.

PROBLEMS IN RADIOACTIVE DATING

As radiometric dating methods were improved, many complex problems were encountered. Most of those problems are beyond the scope of this book, but a few examples follow.

If a sample of rock happened to contain extra lead that had not been produced by the disintegration of uranium, the age calculated for the rock would be greater than its actual age. Fortunately, the lead produced by the decay of uranium was found to be an isotope different from ordinary lead. Consequently, the ordinary lead in the sample could be distinguished from the daughter product by means of the mass spectrometer and excluded from the calculation of age.

A more troublesome source of error was the fact that uranium does not change directly into lead, but changes through some fifteen different elements or isotopes on the way as shown in Fig. 4-2. At one point the disintegrating uranium becomes radon, a radioactive gas, for a short time, before decaying into a new solid, the metal polonium. If an important amount of radon gas were to escape from the crystals of the rock, there would be less lead formed at the end of the process. The measured age would again be incorrect, this time less than the actual age.

Scientists also discovered that there are often two isotopes of uranium present in rocks, each of which decays at a different rate and into a different isotope of lead. While this discovery complicated matters at first, it ultimately proved to be very useful. Using the two isotopes of uranium, and their respective daughter products, two separate measurements of age could be made from the same sample of rock and checked against each other. It was like using two stopwatches to time the same athletic event. With such a check, the scientist could be more confident about the measurement of time, especially if the two measurements were in close agreement.

Uranium has another serious shortcoming in that it is

found in earth rocks generally in very small quantities—about two atoms of uranium per million atoms of rock, on the average. Also, the kinds of rocks that contain enough uranium (or thorium) to permit this method of dating are severely limited. Besides the escape of one of the intermediate decay products—the gas radon, as described earlier—loss of the final daughter product, lead, has also plagued users of the uranium system with inaccurate results.

Consequently, although the uranium-lead method was the most important dating process for a time, it has now been largely replaced by other methods that use other radioactive elements that are found in larger quantities and in more kinds of rocks. These include natural elements such as potassium, rubidium, samarium, and iodine. Radioactive isotopes of elements such as aluminum, argon, chlorine, hydrogen, and sodium that have been created by the bombardment of other elements by cosmic rays are also used.

Another kind of dating uses the radioactivity in the rock in a different way. It measures the damage caused by the two fission fragments of the uranium nucleus when they fly apart with great energy. They travel only one one-thousandth of an inch, but they knock other atoms from their normal positions to make a narrow track through the rock. Under a microscope, these tracks in a polished sample can be counted. If the amount of uranium in the sample is known, the number of tracks reveals the age of the rock. This method, called "fission-track dating," is unique in that ages from 100 years to 4,500,000,000 years (the age of the earth) can be measured.

Thus, the rocks of the earth contain many radioactive clocks that have been recording the time of various geological and life processes since the earth was formed. These clocks can give us accurate, reliable dates if we learn how to read them. Although radiometric dating is a difficult science that strains the very best technologies of physics and chemistry, it has made available to science, for the first time, clocks that run at constant, known speeds.

In contrast, most of the geologic processes that have changed the earth, such as erosion, deposition, upheaval of mountains, cooling of the earth, salting of the oceans, etc., do not take place at a constant rate over the years. As we have seen earlier, trying to use these processes to measure time is as hopeless as using a clock that sometimes runs faster or slower, or sometimes stops for a while, without our knowing it.

RADIOMETRIC CLOCKS ARE VERY RELIABLE

But the rate at which atoms of a radioactive element throw off atomic particles and change into atoms of another element is the same no matter where the rock is located or what happens to it over the ages. Whether the rock is hot or cold, wet or dry, lying on the surface or buried deep in the earth, turned upside down or smashed into pieces, its radioactive atoms decay at the same rate. It is impossible to imagine any kind of "clock" that could be frozen, boiled, smashed, or even ground up into tiny pieces, yet would continue to tick off time with perfect accuracy. But a radioactive element can do just that.

Even melting the rock will not destroy or stop the clock, although it will "reset" the clock back to "zero." This effect is important if we are to understand just what age is measured by radiometric dating. Melting the rock usually clears away the decay products that have already been produced. When the rock cools and solidifies, the elements can no longer move around, but are again frozen in solid rock. Once again the decay (daughter) products begin to accumulate around the parent element as the clock ticks on.

The radiometric clock, then, starts at the time the igneous or metamorphic rock solidifies. It does not tell us the age of the earth, or even the age of the rock itself, but only the time since the last melting.

FINDING THE AGE OF SEDIMENTARY ROCKS

Consequently, most sedimentary rocks (that is, rocks that have been formed from the weathering and erosion of igneous and metamorphic rocks, and subsequent consolidation of the resulting sediments) cannot be directly dated by these radioactive methods.

The processes of erosion, sedimentation, and forming new rock (except when formed by melting) do not "reset the clock." The clock keeps running all through these processes, ticking away in each grain of sand, measuring the time back to the last occasion on which it had been melted. The age of old grains or pebbles that have been included in the formation of new sedimentary rock can be measured by radiometric dating; the age will not be that of the new rock but the date of solidification of the older original rock from which the pebble came. Such a measurement would thus tell us only that the new sedimentary rock was considerably *younger* than the age measured for the older material from which it was made.

It happens that after some sedimentary rocks are formed, they are buried deep in the earth, perhaps under thousands of feet of other soil and rock. Such rocks (or sediments) may be melted by the pressure and heat in the earth's interior and changed into a new kind of rock called metamorphic rock. Because such melting resets the clock, radiometric dating of such rocks will give the time elapsed from the date when the metamorphic rock had cooled down. We would know, too, that the sedimentary rock from which it was made was *older* than the age obtained—perhaps millions of years older.

Although it is frustrating that radiometric dating will not ordinarily give the date for the deposition of a sedimentary rock, it may provide a rough time frame for the geologic events in that area. Such information is always far better than nothing, and may, when used with other information, be very useful indeed.

Fortunately, there is yet another geologic process that provides considerable help in dating sedimentary rocks. Molten rock from deep in the earth is often forced upward under great pressure. It squeezes through cracks and fissures in any rock that gets in its way, finally cooling and solidifying in these places. You can often see such "intrusions" where a highway cuts through a hillside, as streaks of different colored rock in the main formation. If the intruding molten rock, or lava, cuts across the layers of other rock, it is called a dike; if it flows between the layers, it is called a sill.

Radiometric dating of this molten lava tells us when it forced its way into the older rock to solidify as dikes or sills. Although this date does not tell us the age of the older main formation, it does tell us that this rock must have been formed before this date.

As we have seen, although sedimentary strata in the geologic column cannot often be dated directly, dates can be found that are very helpful. The presence here and there of metamorphic (melted and recrystallized) rock, lava intrusions, and layers of volcanic ash or lava may give the geologist many dates in the column. From such dates, and other evidence, geologists have gradually fitted a time scale to the geologic column that now tells us *when* the major geologic eras occurred and *how long* they lasted. Not exact dates, by far, but accurate enough for many purposes.

THE QUEST FOR THE AGE OF THE EARTH CONTINUES

But the known geologic column does not extend far enough back in time to give us the date of the origin of the earth. No rock has yet been found that is believed to be as old as the earth itself. Among the oldest earth rocks yet discovered are samples of uranium ore from Canada and Africa, which dating shows to be almost 3 billion years old. Another rock, found in Greenland in 1971, was measured to be 3.7 billion years old. Later, a metamorphic rock found in Canada was measured to be 3.96 billion years old. And zircons (a brilliant blue-white gem) in a sedimentary rock from Australia were found to have an age of 4.3 billion years.

Scientists had to assume that the earth is older than all of these rock samples, but how much older? The puzzle of the earth's age was not yet solved. Somehow, somewhere, rocks that are as old as creation had to be found to solve that puzzle.

5 CLOCKS IN SPACE ROCKS

THE SEARCH FOR THE OLDEST ROCK

If you wanted to find a rock that had not been disturbed since the formation of the solar system, where should you look? Not on the earth's surface, for sure, now that we know that violent processes have churned the earth's crust to depths of many miles and reset the radiometric clocks of crustal rocks again and again. We could look in the earth's "basement," below the crust where no erosion or deposition has occurred. But no mining or oil drilling has yet penetrated to that great depth.

There's only one other place to look: in outer space. There must be rocks out there somewhere that have not been changed by weathering, remelting, or any of the other slow but violent processes that we know take place on earth. Their radioactive clocks have probably ticked away undisturbed since they first solidified from the original "star stuff" from which the sun and the planets were made. Such rocks—if we can get our hands on them—should be able to tell us the age of the solar system.

Fortunately there are such rocks in space, and fortunately too, we already have a good supply of them right here on earth. They are meteorites, pieces of rock weighing from a few ounces to many tons. While traveling through the vacuum of space from some distant galaxy, or more likely, from some distant part of our own solar system, these pieces of debris came by chance into the path of the earth and plunged into its atmosphere. Three different words are used for these chunks of rock from space. While they are traveling in space, they are called meteoroids. During their fiery entry into the earth's atmosphere, they are called meteors. Those that survive to reach the surface of the earth are called meteorites.

Meteoroids the size of grains of sand are small enough to burn up due to friction in entering the earth's atmosphere. But the larger ones are able to survive and reach the surface of the earth. These samples of cosmic materials, their surfaces pitted and melted by the frictional heat of entry, have been found in large numbers all over the earth. Meteorites of various shapes and sizes can be seen in most of the large museums of the world (Figure 5-1). Meteorites are roughly divided into three types according to their composition: "stones," which consist mostly of stony matter; "irons," which consist mostly of iron and nickel; and "stony irons," which consist of large amounts of both materials.

Dates From Meteorites • When scientists used radiometric dating methods to measure the age of meteorites, the results were astonishing. Most of the dates tended to cluster around the same age—approximately 4,600 million (4.6 billion) years. With this discovery, the age of the oldest rock yet discovered had risen by another *billion* years. Is 4.6 billion years the age of the solar system? If so, is it also the age of the earth? The answer to those questions was not yet clear.

Figure 5-1. Meteorite found in Mexico. The size is about 10 cm. It consists of approximately one-half metal and one-half the mineral olivine.

The age tests on meteorites brought some new mysteries. Some younger ages for meteorites were measured that puzzled scientists for a while. But they later found evidence that these meteorites had been exposed much earlier to heat or shock, probably from bumping into something at very high speed. The resulting heat melted them enough to change the character of their minerals—and to reset their clocks.

Measurements were also made on some meteorites by another dating method—one that measures the length of time they have been exposed to the cosmic radiation of outer space. Many stony meteorites surprisingly showed "exposure ages" of less than 40 million years. These ages were much less than the ages of iron meteorites measured by the same method. Could these stony meteorites have been hidden from cosmic radiation during all these billions of years since the formation of the solar system? If so, where?

One Source of Meteorites: Asteroid Collisions • Scientists now believe that these meteorites may once have been part of larger chunks of debris, such as asteroids—those thousands of small "planets" and pieces of rock that circle the sun between the orbits of the planets Mars and Jupiter. In collisions that took place between asteroids tens of millions of years ago, chunks of newly shattered rock from inside the asteroids became exposed for the first time to the strong cosmic radiation that is everywhere in space. The exposure time we measure dates back to this collision. (Perhaps it was this same collision that threw the meteoroid into a new orbit that eventually brought it into a collision course with our planet Earth.)

This surprising explanation for the birth of meteoroids has been checked in another way. The orbits of some meteors were calculated from photographs of the track of their fiery plunge into the earth's atmosphere. These calculations suggested that the meteoroids were probably on an orbital path that passed through both the orbit of Earth and the "asteroid belt" between the planets Mars and Jupiter. After being knocked out of the asteroid belt and into these new orbits by an asteroidal smashup, these new meteoroids may have circled the sun in their new orbits many millions of times before colliding again—this time with the atmosphere of planet Earth.

Measuring the ages of meteorites from space thus has given us not only strong evidence as to the age of the solar system but also a new understanding of the origin and history of meteorites themselves. The date that is now generally accepted for the age of the solar system came from the ages measured for the meteorites from space: 4.6 billion years. This is several hundred million years older than the oldest rock yet found on earth. But then, we have surmised all along that the earth must be much older than that rock.

DATING THE MOON BY ITS CRATERS

Even before the Apollo moon landings, which began in 1969, scientists had believed that the moon could somehow tell them the age of the solar system. Astronomers had tried earlier to estimate the age of the moon by examining and counting craters on its surface. They found that relative ages of surface features, especially craters, can be estimated by examining their detail and sharpness of outline, and by comparing one crater with another.

At almost any place on the moon there are a variety of craters—some sharp-edged and new-looking, others that look very old and battered (Figure 5-2). On the photographs small, worn craters can be seen everywhere mixed with craters of sharper outline.

Because there is no atmosphere on the moon, there is no "weathering" by wind and water as we know it on earth. Instead, it is meteorite impact over billions of years that erodes and chips away the moon's surface. The older

Figures 5-2a, b. Moon's surface photographed from lunar orbit, showing craters of various ages, sizes, and densities

the crater, the more often have meteorite strikes chipped it away and softened its outline.

Dating craters on the moon is a little like dating strata on earth—it is often easy to tell which feature is older or younger, but very difficult or impossible to even estimate actual ages with useful accuracy.

Some scientists attempted to learn the actual ages of various areas of the moon's surface by counting the number of craters (from photographs like those in Figure 5-2, made from space) and then applying an estimate of the number of meteoroids that might have struck the moon during each million years of its long history. These efforts face the same old questions that have plagued all dating methods: How fast is the clock running? And has it been running at the same speed all the time?

From a count of the craters it was obvious that the maria, or "seas," on the moon have fewer craters and therefore might be much younger than the highlands, or mountainous areas. The maria were postulated to be vast lava flows that took place long after the moon was formed. Their age was estimated, based on the scarcity of craters, to be only 200 to 300 million years.

NEW EVIDENCE: ROCKS BROUGHT BACK FROM THE MOON

One of the main goals of the Apollo missions to the moon was to investigate the moon's age and origin. Moon-bound astronauts were given training as geologists, and spent most of their time on the moon exploring its surface and choosing samples of rocks and soil to take back to earth for analyzing and dating. In all, about 380 kilograms (170 pounds) of soil and rocks were brought back from the surface of the moon. As with earth rocks, the scientists had to be sure that they knew what they were measuring when they dated a sample of moon rock.

Also, as with earth rocks, tests of moon rocks may not give the true age of the rock, but rather the time since the "clock" was last reset. As we have seen earlier, this is usually the date of the most recent melting. Although the moon has no weathering processes like those on earth, and therefore no sedimentary soil, the soil on the moon is *not* undisturbed. The moon's rocks have been broken down into soil and that soil turned over, or "gardened," over a long period of time by two processes: temperature changes (between lunar day and lunar night) and meteoroid impact.

As it turns out, meteoroid impact, which can suddenly splash out new craters or slowly chip old ones into ruins, can also make new rocks out of old ones and reset their radiometric clocks. The shock of the impact and the heat energy produced in the impact can alter the structure of the rock and reset its clock. The age measured for such an altered rock will be less than the age of the original rock.

New Dates in the Moon's History • With careful attention to these and other problems, scientists measured the ages of rock samples from many areas of the moon. Rocks from the maria—those lunar "seas" that were earlier believed to be young lava flows—have proved to be much older than the predicted 200 to 300 million years. The measured ages were around 3,300 million years. Why should such ancient lava flows have so few meteoroid craters and look so young?

Scientists now believe that massive eruptions of basalt (a kind of lava) flowed out over the lunar lowlands perhaps 500 million years after the formation of the moon, then stopped more or less suddenly about 3.3 billion years ago. The fact that these newer lowland areas are not heavily cratered tells us that the moon must have been heavily bombarded by meteoroids only during its very early history.

Probably the heavily cratered areas we now see in the highlands of the moon were formed when the moon collided with huge chunks of rock and debris that were left in space when the moon and earth were formed. After the moon had swept up the debris in its region of space, the bombardment decreased greatly. It was at about this time that the new maria lava flows were taking place. The reduced meteoroid bombardment produced fewer craters over the following 3 billion years, thus leaving these newer maria areas looking much younger than their actual age.

We can surmise from this information that the earth, too, must have been bombarded by large chunks of space debris in its early life. However, most of the damage has been totally obliterated by billions of years of weathering of the earth's surface. Only a few craters remain, the best of which is the famous Meteor Crater in Arizona.

We can see why efforts to learn the moon's age by counting its craters has been a failure: the history of the moon's surface is much too complicated. But as so often happens in science, these efforts led to better understanding of other things, such as the history of the moon's craters and other surface features, and of meteoroids themselves.

METEORITES AND MOON ROCKS GIVE AGE OF EARTH

Rocks gathered by Apollo astronauts from the lunar highlands gave much older dates than the maria—up to about 4 billion years. At least one lunar rock has given a date of 4.6 billion years, the same date given by many meteorites found on earth.

Rocks from space—both meteorites and moon rocks—thus give us strong evidence that the age of the solar system is about 4,600 million (4.6 billion) years. Someday we may find a rock on earth that has been undisturbed since it was first formed. We may do this perhaps by boring into the original undisturbed rock far below the earth's outer crust. It seems likely that such a rock would tell us the same age, because it is probable that earth, moon, sun, planets, and meteoroids were all formed at about the same time.

THE AGE OF THE EARTH—THE NEW VIEW

During the first half of this century, our earth grew older by about a billion years as each decade passed—or so it seemed. As we have seen in earlier chapters, the age generally accepted for the earth was 100 million years in the early 1900s. Then the new science of radiometric dating began to show ages for various rocks from earth and space that were much older: 700 million years, then upward to 2,000 million (2 billion) years, then to more than 3,000 million (3 billion), and finally, up to about 4,600 million (4.6 billion) years.

THE NEW AGE OF THE EARTH

These measurements tell us that the earth is more than forty-five times older than it was believed to be in 1900, and nearly a million times older than the age deduced from Scriptures by Bishop Ussher over 300 years ago. So many different experiments have now agreed upon 4.6 billion years as the age of the earth that it seems unlikely that the accepted age will increase further.

Similar increases in age were wrested by geology over the years from its study of strata and the structure of the earth. For example, if the total thickness of strata for each of the geological eras since the first of the Paleozoic era were combined in a single column, the total height of that column would be nearly 450,000 feet, or about 85 *miles*. The amount of time required to deposit strata some 85 miles deep is difficult to imagine, let alone estimate. Very early estimates that were made gave an answer of only about 26 million years. This estimate may have fitted fairly well the accepted idea of the earth's age at that time (100 million years), but such a period was in fact far too short for such an enormous amount of erosion and deposition to have taken place.

A later estimate, based upon a more realistic view of rates of deposition of about 1 foot of strata per 1,000 years, gave a time period of 500 million years, back to the beginning of the Cambrian era. This answer fits the ages of rocks formed in those intervening ages, as now given by radiometric dating. (We know little about the rock layers that lie below those of the Cambrian era, except that their age probably stretches back another 4 billion years or more.)

During all these billions of years, high places on the earth's surface have been weathered away and deposited in the low spots, then pressed into rock, heaved up as high places to be weathered away again. These processes have churned the earth's surface to such a depth that the undisturbed mantle of the earth has never been seen. Not even the deepest oil well yet drilled has reached below the churned layers of the earth's crust into the undisturbed "basement rocks" that must lie below.

THE GEOLOGIC COLUMN

Scientists in all branches of science attempt not only to discover the principles and laws governing the phenomena they are studying but also to systematize their findings. In the new science of geology, this meant classifying rocks as to how, where, and when they were made.

Over 300 years ago these efforts began with the principle of superposition of strata, described in Chapter 3. Over the years since, geologists have slowly sorted the layers of the geologic column into about 250 "formations," divided these into about thirty-five "epochs," gathered these epochs into about a dozen "systems," or "periods," and grouped these periods into four geologic "eras."

For the purposes of this book, it will be adequate to examine briefly the four eras, the dozen periods, and a few of the recent epochs, and to assign the presently accepted time frame to these divisions.

The Precambrian Era • Starting at the bottom of Figure 3-2 with the formation of the earth's crust about 4.6 billion years ago is the Precambrian era, the longest era, probably more than 4 billion years in length. Life on earth began at some time during this era and probably existed for a very long time in simple forms such as algae and tiny microscopic organisms. Later, simple plants and soft-bodied animals such as worms developed. None of these very early life-forms left reliable fossil evidence of their existence in the Precambrian rocks, except for some markings that are believed to be worm burrows and worm tracks and some other vague impressions that may have been made by algae.

The Paleozoic Era • The beginning of the Paleozoic era, the era of ancient life, about 600 million years ago, is fixed at the first appearance of hard-bodied sea organisms such as brachiopods, a kind of ancient shellfish, and trilobites, a tiny, ancient ancestor of the horseshoe crab. The first system of rocks for this era is named Cambrian, and contains the first fossils found for these creatures. The name Cambrian is taken from ancient tribes that lived in the area of Wales where the rocks formed in this period were extensively studied early in the nineteenth century.

A later period of the Paleozoic, the Carboniferous, is more familiar to us as the period of lush growth of land plants in which most of the world's coal (carbon) was laid down. A still earlier period, the Permian, was named for Perm, a former province of Russia, where the rocks from this period were first studied. The Permian Basin in West Texas was named for this period. Here major oil fields tap huge reserves of petroleum in rocks laid down between 225 and 280 million years ago. In strata laid down during the upper (later) part of the Paleozoic era, fossils of more-advanced forms of ancient life—fishes, spiders, and amphibians—are found.

The Mesozoic Era • Following the Paleozoic is the Mesozoic era, the time of middle life. In strata laid down during this era, from 225 to 65 million years ago, fossils of flying reptiles, the first birds, and those huge reptiles known as the dinosaurs are found.

The Cenozoic Era • The next era, dating from 65 million years ago to today, is the Cenozoic, the time of recent life. Strata laid down in this era contain fossils of life-forms that are increasingly modern: mammals of all kinds, insects, birds, monkeys, apes, and finally humans. The upper (later) period of this era is called the Quaternary (fourth). It is this period, starting at about 2–2½ million years ago, in which the human race developed.

The dozen periods of these three eras are divided into about thirty-five epochs, a few of which are shown at the top of the chart. The Pleistocene epoch was the time of the Ice Age, in which ice sheets, or glaciers, moved down from the north and receded again many times before retreating to their present positions covering most of Greenland, Iceland, and other Arctic islands. The Holocene epoch is only 10,000 years old and covers the postglacial period (or period of moderate glaciation) and the beginnings of human civilization to the present time.

The exciting thing about the geologic column is that it is not just a textbook diagram or a theoretical idea. The column is an actual pile of strata that exists and can be examined. From the Mesozoic and Cenozoic eras, for example, there are some 250 formations that are thick enough and distinctive enough to be called separate units of the geologic column. Not all of the strata in the column exist at any one spot on earth. At any one place, there are often some layers missing because not all strata deposits were laid down everywhere at the same time. Or, the missing layers may have been eroded away again. But when the layers are correlated and brought together, they can be read "page by page" like the pages of a giant history book, using fossils, dating methods, and the other tools and principles of geology.

THE GRAND CANYON

Nowhere on earth is there a grander, more breathtaking place to read this history than in the strata of the Grand Canyon in Arizona (Figure 6-1). The Grand Canyon is a great rugged gash in the earth's crust, wrapped around 277 miles of the Colorado River. As much as 18 miles wide and over a mile deep in some places, the Grand Canyon is by far the most startling work of nature found on the American continent.

How the Canyon Was Formed • Many ideas were expressed in the early days of geology as to how such a gigantic chasm could have been formed. The most obvious idea was that a wide crack had somehow opened in the earth's crust. Another idea was that the canyon was a cave

Figures 6-1a, b. The Grand Canyon provides an unparalleled view of both layering (strata) and erosion.

that had long ago been hollowed out by running water. Then the "roof" fell in, leaving a canyon with rubble and a river at the bottom.

All such ideas have long ago been discarded. Careful study of the canyon and the surrounding region has shown that the canyon was cut, worn, and carved by the Colorado River out of solid rock over a period of about 10 million years. The canyon is still being dug wider and deeper today as the river scours and grinds away at its channel with the soil and rocks washed down the canyon walls by rain and small streams.

The 10 million years that it took the river to dig the canyon is not of as much interest to us today as the previous 2 billion years during which other rivers deposited the mile-thick slab of sedimentary rocks in which the canyon was later carved. The layers in this mile-high pile of sediments are pages of the history of that period (Figure 6-2).

The most recent page is the one you walk on when you visit the canyon and stand at the edge to enjoy the breathtaking view. The oldest page is the hard, dark, wrinkled ancient layer at the bottom. Between the two lie 600 million years of the earth's history.

Below the oldest rocks at the bottom of the canyon lie still more rock that is estimated to be 5 to 6 miles thick—half a dozen times the thickness of the strata exposed in the canyon walls. We know almost nothing about this buried rock, and because it is so inaccessible, it seems questionable whether we shall ever learn very much.

Because the geological history of the Grand Canyon area is long and complex, covering it in detail is beyond the scope of this book. (See For Further Reading for other books on this topic.) But we can trace the main events and then turn quickly through the "pages" of the visible strata.

Several billion years ago, the 5- to 6-mile stack of sediments that became the dark Vishnu schist (the rock now at the bottom of the canyon) was laid down. Buried

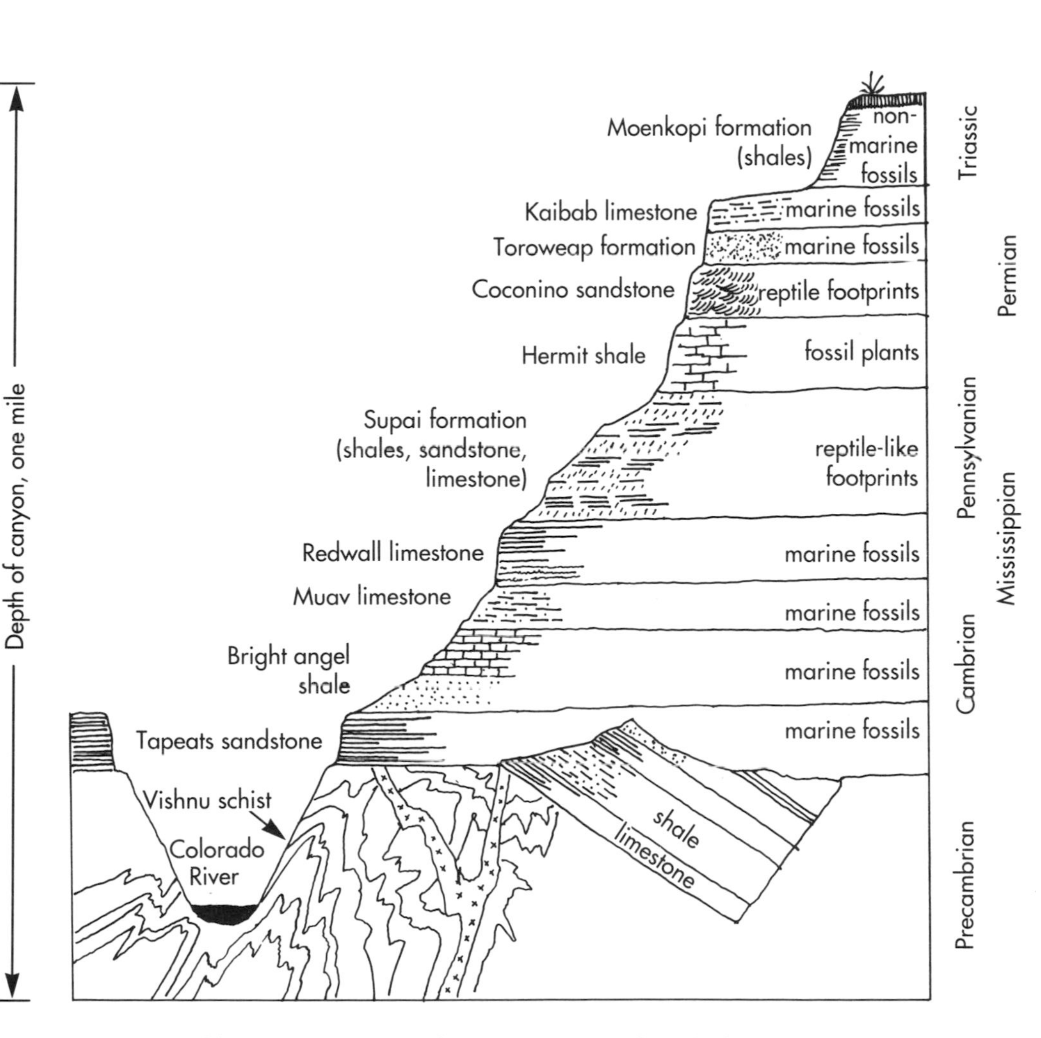

Figure 6-2. Geologic cross-section (schematic) of the Grand Canyon of Arizona

under still more sediments to a depth of 10 to 15 miles, these sediments were twisted and squeezed and mixed with lava and cooked into the hard metamorphic (changed) rock mass whose top we now see peeping out of the bottom of the canyon.

This deformation of the crust squeezed up mountains at the earth's surface, which in time were entirely eroded away by wind and water. This land again sank beneath the water, and new sediments, now called the Grand Canyon series, perhaps 2½ miles thick, were laid down. These sediments were later uplifted, broken, and tilted. A new range of mountains again rose above the surface, only to be eroded away once more, this time almost down to the Vishnu schist below. Only a few wedges of the 1,200 million-year-old Grand Canyon series remain today near the bottom of the canyon, as we shall see in a moment.

Around the beginning of the Paleozoic era (the era of ancient life), 600 million years ago, the flat eroded plain sank once more beneath the water, and the Paleozoic strata which we now see in the walls of the Grand Canyon began to accumulate. The deposition process took about 400 million years (all of the Paleozoic and some of the Mesozoic era) to complete. Once more the land rose, and erosion completely stripped away the nearly 2-mile-thick top layers of rock, leaving on top the Kaibob Limestone layer on which we walk today at the rim of the canyon. About 10 million years ago, as the land continued to rise, the Colorado River began to carve out the 277-mile-long trench that we call the Grand Canyon.

The River Grinds Out the Canyon • Although you might reasonably assume that the river cut *down* through the canyon from the level of the rim we see today to its present channel, this is not in fact what happened. Instead, the land rose about as fast as the river channel was worn downward. The river therefore did not slice down through the land like a knife moving through a layer cake; rather, the cake moved upward to be sliced against an almost stationary knife.

We don't know whether the land around the canyon is still rising or not, so slow are such geological processes. But we do know that the river is still carving the canyon deeper, carrying downstream some 500,000 tons of debris and sediment per day.

We might pause here to ask: just how much material is 500,000 tons? A medium-size ocean freighter carries 10–20,000 tons; hence, the river carries away 25 to 50 shiploads of soil and debris every day. At flood stage the river has carried as much as 27 million tons, or some 1,200–2,500 shiploads, per day. Much of this is fine dirt and sand, while some is in the form of rocks and boulders that bump and crunch their way along on the rocky bottom, grinding the riverbed—and the whole canyon—deeper at the rate of a foot or so every 20,000 years.

One of the reasons why the Colorado River grinds hard on its rock-lined canyon and carries off so much material is its rapid drop. The river's channel slopes downward nearly 8 feet for each mile of travel. This is nearly twenty-five times as steep as the slope of the bed of the Mississippi River, which is slow and sluggish as rivers go.

Other weathering processes are still carving the canyon far above the present reach of the river. Wind, rain, sun, the processes of freezing and thawing, and even the activities of plants and animals, are constantly chipping away at the exposed rock. Pieces of rock and soil regularly fall downhill toward the river (sometimes in huge chunks), where they may eventually be swept downstream to leave the canyon a little wider and deeper.

READING THE PAGES OF THE GRAND CANYON'S HISTORY

As we noted earlier, the geologic column represents actual rock formations. If we were to climb up one of the trails that is notched into the steep walls of the canyon, we could examine on our way many of the rock layers of the geologic column.

The "Basement Rocks" • If we were to stand at the level of the river, the dark rocks towering some 1,500 feet above us would be the Vishnu schist. This hard gnarly-looking rock was originally a sedimentary layer but has been crystallized by heat and pressure into a solid mass of metamorphic rock. The Vishnu is a Precambrian rock, more than 1,700 million (1.7 billion) years old.

Primitive Life-Forms • Just above the Vishnu, we would find in some places the tilted blocks of the Grand Canyon series, all that is left of that pile of Precambrian sediments that was once 2½ miles thick. The first layer of the series atop the schist is the Bass limestone. It contains no real fossils but some "fault casts" of what may have been primitive corals and other prints that may have been left by algae and primitive plants. These animals and plants were among the only living things on earth at that time. These rocks, too, were laid down in the Precambrian era, some 1,200 million (1.2 billion) years ago.

More-Advanced Life-Forms • The first layer (above the Vishnu) of the newer layers that form the main walls of the canyon is the Tapeats sandstone, a Cambrian rock laid down near the beginning of the Palaeozoic era 600 million years ago. Fossil worm tracks and trails of trilobites tell us that crawling, creeping animal life was beginning to appear in the seas (or waters) of that era. Above the Tapeats is the Bright Angel shale, which contains more worm tracks and numerous fossils of more advanced life, such as seaweed, brachiopods, and trilobites. These creatures were the kind of life that lived on the earth 530 million years ago.

Life Crawls Out of the Sea • The layer above, the Temple Butte limestone, contains few fossils, while the next layer, called the Redwall limestone, yields a variety of corals, shellfish, and extinct snail-like animals from the early Carboniferous period. The Supai formation, just above, contains no marine fossils, which suggests that it may have been a floodplain not constantly under water. It contains mud cracks, raindrop prints, plants, and tracks of land animals that walked or crawled on its then muddy surface 285 million years ago. This formation was laid down in the latter part of the Carboniferous period, the time when coal deposits were also being formed in many areas of the world.

Advanced Animal Life • The younger strata from here up shows the advance of life on the land now, with the Hermit shale yielding freshwater fossils, animal footprints, ferns, leaves, insect wings, and cone-bearing plants. Just above is the Coconino sandstone, which seems to have been laid up as dry dune sand. It contains tracks and footprints of primitive reptiles and amphibians. At the top of the canyon in the Kaibab limestone are more sea deposits containing a variety of shells (from some eighty kinds of invertebrates), corals, and sponges.

The Missing "Pages" • Here the history of the earth in this particular "book" abruptly ends, its later "pages" having been torn out by the erosion that lopped off the layers above—layers that once contained the history up through most of the Mesozoic era. But fortunately, layers from that era, with their fossils of still more advanced creatures, exist elsewhere in the world, and they have told us their part of the history of life and the earth's crust.

The history of both the earth and its living things is far longer, far more complex, and far more fantastic than could ever be imagined before paleontology (the study of fossils), geology (the study of the earth), geologic dating, and related sciences, showed us how to read it. In the next chapter we will return to the problem of dating events that occurred relatively recently. We will examine the discovery of dating methods that made it possible to learn how long ago a stone tool or clay pot was made, or how long ago a packet of seeds, a wooden coffin, or a human skeleton was placed in a grave. New dates for such events have made possible a more accurate history of the human race in prehistoric times.

7 DATING BONES, CHARCOAL, AND POTTERY

As scientists began to find out that the earth we inhabit is the product of a very long geological history, they began to realize too that the human race is similarly the product of a long biological history. Slowly, the world began to recognize that the human race must be much older than the few thousand years that was first believed. As in the case of earth's history, the beginnings of the human race began to stretch back farther and farther into the past as more knowledge was gained.

AN ABUNDANCE OF ARTIFACTS

The human race has left behind a great many records of past life and past cultures in graves, pottery, clay figures (Figure 7-1), tools made of bone, wood, and stone, and in dwellings (especially caves) that contain seeds, the remains of campfires, and the bones of eaten animals. These artifacts and other archaeological evidence have made it possible for scientists to understand how early people lived, how they were clothed, what they ate, how and when they

Figure 7-1. This Mayan ceramic figure from Campeche, Mexico, dates from the seventh century.

developed tools, and even their ideas on religion. By analyzing such evidence, it has been possible to trace the upward progress of early humans toward civilization. But what was missing during the early period of this work was some way of determining the *ages* of artifacts in order to learn *when* important events or important developments took place.

Tree-ring dating, as we have seen in Chapter 1, made it possible to date dwellings back a few thousand years, but not earlier than the dawn of recorded history. On the other hand, radiometric methods for dating rocks were found to be useful for dating only events that occurred more than about a million years ago. Thus, there seemed for a while to be no "clocks" or "calendars" to date artifacts in the period between a few thousand years ago and a million years ago. This large gap left the archaeologist without methods for obtaining accurate dates during the most interesting period of the development of the human race—the prehistoric period during which early people were changing from prehumans to nomadic hunter-gatherers and later to farmers and village dwellers.

But radioactivity finally came to the rescue of the archaeologist when a new kind of radioactive clock was discovered. The basis for this clock was the startling discovery in the 1940s that all living things are radioactive!

THE DISCOVERY OF RADIOACTIVE CARBON

Early research in radioactivity had shown that a radioactive form of carbon, called carbon ^{14} or radiocarbon (chemical symbol, ^{14}C), could be produced in the laboratory when nitrogen was bombarded with neutrons. An American scientist, W. F. Libby, reasoned that conditions existed high in the earth's atmosphere for the creation of ^{14}C by natural processes. It was well known that high-energy nuclear particles, known as cosmic rays, stream to the earth from outer space. These particles collide with atoms of gases in the earth's upper atmosphere, smashing them to release other nuclear particles, including slow-traveling neutrons.

These neutrons, Libby reasoned, could collide with, and be captured by, atoms of nitrogen in the earth's atmosphere to change these atoms into carbon^{14}. These atoms would be expected in turn to react with atoms of atmospheric oxygen to form carbon dioxide (CO_2) molecules. Because living things build their bodies from carbon atoms taken from atmospheric carbon dioxide, Libby further predicted, carbon^{14} would be found in the leaves and stems of plants and in the bodies of animals that eat those plants.

No one at that time had ever detected carbon^{14} in living things. Libby calculated that the amount in the atmosphere would be extremely small—only one atom in a thousand million normal carbon atoms. He then conducted an investigation which showed that carbon^{14} is indeed present both in the atmosphere and in all living things. Because carbon^{14} contains in its atomic nucleus two more neutrons than natural carbon (carbon^{12}), it is "unstable" and radioactive; consequently, all plants and animals are radioactive.

This may come as a surprise, but *you* are radioactive, too! Your body, your clothes, this book, all contain radioactive atoms of carbon^{14} that are throwing off radiation as they slowly but steadily decay into atoms of other elements. When we are in a crowd, our bodies are being bombarded by extra radiation from other people, because they are radioactive too.

It was soon found that the carbon^{14} in living things could be used as a clock to measure the age of bones, samples of wood, cloth, seeds—anything that had once been living and had not been dead for too long.

HOW RADIOCARBON DATING WORKS

How does this clock work? All living things take up atoms of carbon^{14} as part of the carbon with which they build their bodies. Plants do this in the process of photosynthesis, animals by eating the plants. When the organism—plant or animal—dies, the process stops and no more carbon ^{14} is acquired. Because carbon ^{14} is unstable, it slowly decays, with a half-life of about 5,700 years, into nitrogen ^{14}. The nitrogen gas disappears into the atmosphere. After 5,700 years, half of the carbon ^{14} has decayed; after another 5,700 years, half of the remaining half has decayed (leaving only one-fourth of the original), and so on.

We know or can determine the amount of carbon ^{14} originally in the material and the rate at which it decays and disappears. We can measure how much carbon ^{14} is left (which is analogous to the length of the candle left unburned in Figure 3-3, in Chapter 3) by measuring the amount of radioactivity remaining in the sample. With these quantities, we can calculate the age of the sample or, more precisely, the time from the death of the living material to the time of measurement.

CHECKS AGAINST OTHER METHODS

Reliable checks of this method have been obtained by testing samples of material of known age—for example, wood from ancient coffins or funeral boats, seeds, and cloth—all from Egyptian tombs whose dates are known from other historical records. A funeral boat from the tomb of Sesostris III has a historically determined age of 3,792 ±50 years; the age determined from carbon ^{14} dating was 3,700 ±400 years. An Egyptian coffin from the Ptolemaic period, with a historical age of 2,280 years, gave a ^{14}C age of 2,190 years. Although carbon ^{14} dating does not give dates as exact as historians would like, it has provided a much needed time scale for events over many thousands of years of important human history—dates which would otherwise have been uncertain or unknown.

Also, charcoal buried by volcanic eruptions has made it possible to date eruptions that occurred within the last

10,000 years. After 50,000 to 70,000 years of radioactive decay, the amount of carbon^{14} remaining is so extremely small that materials older than this cannot be accurately dated by this method. At about 50,000 years, only 1/5 of 1 percent (or 2 parts out of each 1,000) of the original carbon^{14} remains. The difficulty of measuring such small quantities comes into focus when we recall that the amount of carbon^{14} in living things is only 1 part in 1 billion at its *maximum;* that is, at the death of the organism.

Scientists desperately need to learn the age of charcoal found in ancient campfires that burned around what is believed to be a quarter of a million years ago. But alas, radiocarbon dating is no help here. After this much time, the carbon^{14} has decayed to near zero; in other words, the clock has run down and almost stopped.

OTHER DIFFICULTIES WITH RADIOCARBON DATING

There are many other problems that make radiocarbon scientists' lives difficult—and plague them with wrong answers. Extra carbon can be put into a sample—an old buried bone, for example, can pick up new carbon from plant roots that grow through it. Volcanoes that give off carbon dioxide may locally enrich the atmosphere with ordinary carbon dioxide and thereby reduce the amount of radioactive ^{14}C that is picked up by nearby plants. Some chemicals in seawater can change the amount of ^{14}C that is picked up by sea life and thereby cause error in the measurement of age. And then there is the possibility that the amount of radiocarbon in the atmosphere has not been a constant over the years. (This is the question raised earlier in this book about other dating systems: has the "clock" been running at constant speed?

Samples of wood that have been accurately dated by tree rings or other methods have come to the rescue of the scientists by giving them specimens of known age to check by the radiocarbon method. Measuring the age of such ancient samples has indicated that the amount of ^{14}C in the atmosphere (and in living things) was about the same back to around 3,000 years ago. Earlier than this date, the ^{14}C level was different from the present level, and was probably about 8 percent higher around 10,000 years ago. These differences are believed to be due to changes in intensity of the cosmic rays that produce ^{14}C in the upper atmosphere.

Other investigations have showed that the relative level of ^{14}C in the atmosphere dropped several percent around the year 1900. At about that time there was a large increase in the amount of ordinary carbon dioxide released into the atmosphere, probably as a result of the burning of greater amounts of coal and petroleum (to produce electricity and motive power). More recently, a sharp increase in the amount of ^{14}C has been found; nuclear-bomb testing in the atmosphere has been cited as the cause.

All of these changes and uncertainties, plus the extremely small quantity of radioactive material in each sample, has made radiocarbon dating more difficult and less accurate than historians and archaeologists would like. But despite those limitations, radiocarbon dating still provides a useful clock. With this method, bones, charcoal, seeds, and wooden implements from ancient civilizations can be made to reveal their ages with useful accuracy, provided that they are not older than about 50,000 to 70,000 years.

Radiocarbon dating has done a great deal to untangle the difficult puzzles surrounding prehistoric people, and has sometimes made orderly history out of what was once chaos.

OTHER CLOCKS AND CALENDARS

A charred loaf of bread from Pompeii, the cloth wrapping from an ancient Egyptian mummy, and the linen wrapping from the Book of Isaiah in the Dead Sea Scrolls—apart from the value of dating them accurately, artifacts are themselves astonishing glimpses thousands of years into the past. But the need to date important artifacts is of course inescapable. And what of those artifacts that do not contain carbon or other radioactive materials? They have made it necessary for scientists to search for other clocks and calendars that could be used to find ages.

One of the artifacts that archaeologists have plenty of, at least from digs dating back many thousands of years ago, is pottery. Pottery items have been called the "archaeologist's fossils." But like fossils, pottery can be very frustrating. The scientist may have some fragments of a jug, or even a complete vase or bowl, that tells much about the civilization in which it was made. But the scientist really needs a date to tell how old it is, and manufacturers of those days didn't inscribe dates on their products.

In attempts to solve this problem, some very exotic methods have been developed, one of which is thermoluminescent dating. This method depends upon the fact that crystals of material (pottery, for example) slowly store energy in their electron structure when natural radiation (cosmic rays) strikes them. When such crystals are heated in the laboratory to a glowing temperature, some of the light they emit comes from this stored energy. The older the crystal, the more radiation it has absorbed and the more brightly it will glow. Very careful measurements of thermoluminescence, then, can give a measure of the age of the sample. This very difficult method has given some useful results in dating pottery, bones, teeth, and volcanic ash that is relatively young, geologically speaking. It has also been used with some success as a check against carbon ^{14} dating. Dating of bones that are more than 100,000 years old has been performed with some success.

Scientists have found some chemical clocks that give a measure of age. It has been found that old bones buried in the earth gradually change in various ways, slowly losing or absorbing certain chemicals. Fluorine, for example, can be slowly absorbed from groundwater and stored in the bone.

Solving the Hoax of the Piltdown Man • Fluorine dating was used to date the remains of the so-called Piltdown man. This famous fossil was "discovered" in 1912 in a gravel formation on Piltdown Common near Lewes, England. Fragments of a skull and jawbone were claimed to have been found, along with fossil remains of ancient animals now extinct. A controversy over whether or not these fragments were from a distant ancestor of man from the early Pleistocene epoch (200,000 to 1,000,000 years ago) raged for about forty years. The problem was that the Piltdown man was different in several important respects from other ancient men of that era whose fossil remains had been discovered earlier. Piltdown had a large brain case, no eyebrow ridges, and strangely worn teeth, none of which fitted into the known fossil record of human development. In 1926 the age of the gravel in which the fragments were "discovered" was measured and found to be much less ancient than supposed, which cast further doubt on the fossils.

The controversy was suddenly resolved in 1953 when measurements of several chemicals in the fossilized bone, such as fluorine, nitrogen, and organic carbon, showed that the skull and jaw were not of the same age (and therefore not from the same creature), and that neither was very old. Carbon^{14} dating tests a few years later confirmed these findings, and further investigation showed that the fragments of bone had been stained and altered to make them look old and authentic. Piltdown man was not an ancient member of the human race, but a hoax. (Just who perpetrated this elaborate hoax and why has never been learned.)

Dating Stone Tools • Chemical changes that occur over time to the rock obsidian have given scientists another interesting dating method. Obsidian is a volcanic glass, usually lustrous black but sometimes red or brown. Early people used this material to make knives, arrowheads, and other weapons and tools. When a surface of obsidian is freshly exposed, as when a new tool is formed by chipping, the fresh surface of the rock begins to slowly absorb and combine with water chemically—a process called hydration. The rate at which this hydration penetrates into the glass depends only on temperature and is therefore the same in wet or dry climates. The rate is extremely slow—only a few thousandths of an inch in 50,000 years.

An ancient obsidian tool that has not been abraded or exposed to fire since it was chipped out can be dated by slicing the object and measuring the depth of the hydrated layer that has been growing slowly but steadily thicker since the tool was fashioned.

The various methods for dating materials—ancient logs, carbon materials, rocks, pottery, and all the others—have helped immeasurably to unravel the mysteries of our ancient planet and its ancient life by providing dates for events that happened thousands and millions and billions of years ago.

The greatest puzzle of all still remained: how old is the universe? A special difficulty is that samples of the universe cannot be gathered for measurement. Fortunately, it turns out that the universe is itself a gigantic clock that can be read from afar with the telescope, modern instruments, and a great deal of ingenuity.

8 DATING THE UNIVERSE

In science fiction and in comic strips, we sometimes read about "time machines" that can let people look—or even travel—backward in time. With such a machine, time travelers can witness or even experience events that took place long ago.

Can such a time machine be built? Not really; the past is past, and nobody believes that an event can be reobserved or relived once it has gone by.

THE UNIVERSE AS A TIME MACHINE

But curiously, the universe itself is a sort of time machine which allows us to see events long after they happen. When you look at the sun, for example, you are seeing it, not as it is at that instant, but as it was 8 minutes earlier. The delay is caused by the fact that the speed of light is fixed at 186,000 miles (300,000 kilometers) per second; consequently, sunlight takes about 8 minutes to travel the 93 million miles from sun to Earth. If there is an eruption of flame on the sun (which sometimes happens), we do not know about it until the light reaches Earth about 8 minutes later.

Looking Backward in Time • We look even farther back in time when we look at the stars. The star nearest to Earth, called Alpha Centauri, is 26 trillion (26,000,000,000,000) miles away, so far that its light takes 4½ years to travel to the Earth. (We say that this star is 4½ light-years away, a light-year being the distance traveled in a vacuum by light in 1 year.) When we look at this star, then, we are seeing it not as it is today, but exactly as it was 4½ years ago.

That's the *nearest* star. When we look at a star out at the outer edge of our galaxy (that hazy band or cloud of stars in the night sky overhead that we call the "milky way"), we are looking a distance of 80,000 light-years into space. If we were to see that star explode (which sometimes happens), we would know that the explosion actually took place 80,000 years ago, when the light from that explosion left that star on its long trip to Earth.

That's the nearest galaxy, the one that contains our sun and solar system. Other galaxies are millions or billions of light-years away. We see them too (through a telescope) just as they were those many millions or billions of years ago (Figure 8-1). In earlier chapters we learned that fossils have preserved for us a view of what ancient life looked like. The ancient starlight that took so long to get here is, in fact, a kind of "fossil." It shows how the universe looked eons ago.

The farther we look into space, the farther we look back into time, back into the history of the universe. How far back can we look? Was there a beginning to the universe? If so, how long ago was it formed? And by what process?

In this chapter we will trace briefly the discoveries and events that allow us to literally look back in time to

Figure 8-1. Because the light from the very distant stars that make up this spiral galaxy (NGC 6946) took many millions of miles to reach earth, we see this galaxy as it looked many years ago. (The white dots are "nearby" stars.)

obtain a fuller understanding of the universe—how it was formed, how old it is, and what may become of it in the end.

THE UNIVERSE IS CONSTANTLY CHANGING

If the distant stars were just floating out there in space at those distances, we might have no way to tell how old they are, or when they—or the universe—were formed. But the key fact is that the universe is not static, or standing still, but is rapidly moving and changing. Like the growing tree rings, the burning candle, and the decaying radioactive substance, these changes are "clocks" that can be used to measure time.

Early Ideas of the Universe • The universe is one of those things that becomes more complicated and more fantastic as we learn more and more about it. Our ancient ancestors long thought that the sky was a sort of roof, or dome, over the world. The stars, they believed, were merely points of light attached to or suspended from its surface.

A major step was taken when people began to realize that there is no dome there, but rather a vast void in which float millions and billions of stars, each a sun more or less like ours. Galileo (1564–1642) helped immeasurably when he turned his homemade telescope skyward and found that the moon was not just a light in the sky, but another "world" of rock and dirt, with mountains and craters scarring its surface. He also learned that the hazy cloud of light we call the Milky Way is actually a vast swarm of millions of stars. It was later learned that these stars make up a galaxy, a vast, circular, disk-shaped swarm of stars in which our solar system is located.

As stronger telescopes were developed, astronomers discovered many mysterious, soft smudges of light, but could not tell whether they were groups of stars or only pockets of gas and dust. For a while it was not known whether these formations were inside our own galaxy or far outside it. Later it was learned that many of these smudges of light are actually other galaxies, giant swarms of stars very, very far away. Our knowledge of the universe took another gigantic leap forward when we learned that there are uncountable billions of these galaxies, each containing millions, or even billions, of individual stars.

RADIATION BRINGS US ALL OUR INFORMATION

The only information that we have about the galaxies and the stars is the light and other radiations that come to us through billions and trillions of miles of space. Naturally, astronomers have worked hard to find ways to analyze these radiations. Sir Isaac Newton didn't know it at the time, but he discovered a most valuable tool when he found that light can be broken down into its component colors (called a spectrum) by passing it through a glass prism. The band of colors thus produced is the same phenomenon that we see in a rainbow. Other scientists discovered 150 years later that every material, when heated to glowing, gives off a spectrum that has a unique pattern of colors. Another scientist, a few years later, found mysterious dark lines in spectra reaching the Earth. Further investigation showed that the dark lines were caused by the fact that certain wavelengths of light are absorbed when the light traveling through space passes through gas or dust.

Astronomers soon found that by analyzing the spectra of stars, they could learn what materials (elements) make up the stars, and what kinds of materials are out there in space (as gas and dust) among the stars as well. The new science of "spectral analysis" made it possible to learn how stars produce their energy, as well as how they are formed and how they die. This new knowledge led to ways of measuring the age of various types of stars, which in turn gave new information on the question of the age of the universe.

One of the big unanswered questions was: are the stars stationary, or are they moving through space? Again, it was the light from the stars that gave an answer to that question. We now know that the stars are not motionless in the void of space—they are moving away from each other, and away from the Earth, at enormous speeds!

MEASURING THE MOTION OF THE STARS

The measurements that tell us this depend upon the "Doppler effect," discovered over a hundred years ago by an Austrian physicist, Christian Johann Doppler (1803–1853). This effect is one that we all have observed—or heard—on Earth.

The Doppler Effect • When an automobile or a train passes us with its horn blowing, we hear the horn at a higher-than-normal pitch as it approaches. As the vehicle passes and moves away, the sound of the horn abruptly drops to a lower pitch. This effect is caused by the fact that the sound waves tend to pile closer together *ahead* of a moving source, and to stretch out farther apart *behind* the source. When the sound waves pass our ears, we hear the close-together waves as a sound of higher pitch (frequency), the farther-apart waves as a sound of lower pitch (frequency).

The Redshift • Light is different from sound in two ways, as regards the Doppler effect. First, light travels at 186,000 miles (300,000 kilometers) per second—nearly a million times faster than sound—and therefore can be used to measure only very high speeds. Second, the motion of a light source that produces a change in frequency is seen as a change in the *color* of the light. For a light source that is moving away from us, the light reaching us has been shifted toward a lower frequency and we see that light as more red. If the motion of light source is toward us, the light is shifted toward higher frequencies and we see that light as more blue. In astronomy, this effect is called the redshift.

Astronomers were surprised to find that light from distant galaxies showed a shift toward the red, which should mean that these galaxies are flying away from us. Could this really be true? Astronomers were unsure at first, especially because the measurements showed that the galaxies were moving at speeds as great as 1,000 miles per second! Several astronomers who studied the data thought they saw evidence that the galaxies farther away might be moving away even faster than those close to us.

The suggestion that the universe might be expanding had been made by other scientists before, without proof. Now that solid evidence was at hand, the study of this idea began in earnest.

THE "BIG BANG" THEORY

Using Einstein's theory of relativity, physicist George Gamow and his associates offered the explanation that the universe had begun in a tremendous explosion, which has come to be called "the Big Bang." The universe, they say, has been expanding ever since. This astonishing idea says that the universe began with a tiny, highly condensed, overheated blob of something, probably mostly energy, perhaps no bigger than a single atom. In the explosion fireball, within a few seconds after the explosion, chemical elements began to form, and that process was completed in time equivalent to a day or less. The universe by then contained more matter than energy.

For more than 10 million years, and possibly longer, the universe was in darkness, as the matter created in the explosion flew outward in all directions. The matter began to gather into clumps that became galaxies, and in those galaxies smaller clumps gathered that were to become stars. As the new stars burst into flame, their thermonuclear reactions ignited by the tremendous heat and pressure inside the star, the universe began to twinkle into light.

Scientists tell us that we must avoid thinking of the Big Bang as a bomblike explosion *in* space or *in* the universe. Instead, we must think of it as the beginning of the expansion of the Universe itself, an explosion *of* space at the beginning of time.

Where did that tiny speck of primordial energy come from? Why did it explode to create the universe? How long ago did it happen? We may never know the answer to the first two questions, but we are well on our way to answering the third.

The measurements showing that the galaxies of the universe are flying outward away from each other not only support the Big Bang theory but also give the universe a point of beginning. Knowing from redshift measurements the speeds at which the galaxies are flying away from each other, can we work backwards from the present time to find how long ago it was that they departed the point of the Big Bang? We could use this method—if we knew how far away the galaxies are. For a time, that information was missing.

The Need for Measuring Distances • But astronomers were developing a way to acquire that information. Stars grow dimmer and dimmer as we look farther and farther into the distance. We have long known that the light we receive from a star varies inversely as the square of its distance. A star of the same brightness at twice the distance will shine with only one-fourth the light; at five times the distance, one twenty-fifth the light; etc. The problem is that stars come in all sorts of sizes and brightnesses, and most are of limited use in measuring distances.

A mysterious kind of star, called Cepheid variable, came to the rescue. This star, for reasons that are not entirely understood, puffs up to a larger size and shrinks back to original size in a regular cycle. Its surface temperature changes too, with the result that the star varies in brightness as it changes size. Dim Cepheids were found to change from dim to bright and back to dim again in a few days, their maximum brightness equal to only a few hundred suns. The most luminous Cepheids change more slowly, with periods of 100 days from dim to bright and back to dim, with an average brightness of 10,000 suns.

These characteristics would be of no help at all in obtaining distances if it were not for a lucky circumstance: the brightness of a Cepheid variable was found to be closely related to its period of fluctuation. Astronomers have measured the brightness and periods of some Cepheid variables at known distances from Earth and have been able to chart with useful accuracy the relationship between period and brightness. If we measure the period of a distant variable, we can find from the chart the intrinsic brightness (its actual brightness at zero distance away, as it were) of that star. By comparing this brightness with the *apparent* brightness (as it appears from Earth), we can calculate how far away the star is.

Only one more thing was needed: very bright stars that would provide enough light to permit measuring redshifts for galaxies that lie very great distances from Earth. In the 1960s, astronomers discovered what they needed—mysterious bodies that radiate enormous amounts of energy into space. At first they were thought to be nearby stars, but redshift measurements showed them to be very far away indeed, and far more powerful than originally thought. Though the subject of considerable research, no one yet knows just what they are. Because their output of radio energy is so strong, they were dubbed "quasi-stellar radio sources," or quasars. The presence of quasars in faraway galaxies gives astronomers the strong beacons they need to measure redshifts at those distances.

THE SIZE OF THE UNIVERSE SHOWS US ITS AGE

With Cepheid variables to tell us the distance to a galaxy, and the redshift of quasars to tell us the speed of a galaxy through space, we have the tools to measure not only the size of the universe, but also how long a time it has taken for the universe to expand to that size. This time is the age of the universe! The American astronomer Edwin Hubble combined these methods, along with measurements of speed and distances of many faraway galaxies, to create a simple graph of distance against speed of recession, which came to be known as Hubble's law. After an astronomer has measured the redshift speed of a galaxy, he can use Hubble's curve to determine the distance from Earth to that galaxy.

Measurements of redshifts have showed that there are many distant galaxies that are flying away from us at up to six-tenths of the speed of light—that is, at about 111,000 miles per second. Fewer galaxies have been found that are moving at speeds of up to about nine-tenths the speed of light—167,000 miles per second. The distance corresponding to that speed of recession is about 18 billion (18,000,000,000) light-years, or about 36 billion trillion (36,000,000,000,000,000,000,000) miles. Surely this object must be close to the observable limit of the universe. If we were to find a galaxy with a velocity of 100 percent the speed of light—which is not likely—its distance would, according to the Hubble curve, be about 20 billion light-years away.

The distance to a galaxy in light-years would be equal numerically to the number of years since the Big Bang, *if* the galaxy had been traveling at the same speed since the beginning. But gravitational forces have undoubtedly slowed the speed of expansion since the beginning; hence, the age of the universe is somewhat less, in light-years, than the number given by Hubble's law. With this and other adjustments, the true age of the universe is generally estimated to be between 10 and 20 billion years.

There are still some doubts about such measurements, and the search for a more accurate age for the universe goes on. The quasars are still a mystery, and some astronomers still wonder if the speeds and distances indicated by their redshifts are really true. If some unknown effect is causing an error, then the size of the universe, and the age of the universe, may be either less or greater than is now believed.

LOOKING BACK TOWARD THE BEGINNING OF TIME

The universe is a true "time machine" that boggles the mind. When we look at a galaxy 15 billion light-years away, we are seeing it as it looked when it was very young, soon after the formation of the universe. Quasars, those powerful beacons whose radiations are used to measure redshifts, are not found in near space but only at great distances from Earth. Why is that?

Scientists believe that quasars, whatever they are, are relics of the early eons of the universe and have all disappeared in the "old" universe near us, as we now see it. Quasars can still be seen in the distant universe because we are seeing that part as it was back when it was "young." The quasars we see in the far distant "young" universe are really "not there" now, but have disappeared or changed into something else many billions of years ago.

Would observers in a galaxy of that far universe, peering in our direction through telescopes, see ours as a "young" galaxy with quasars? Yes, they would. The light that they would see today would have left our galaxy 15 billion years ago and would show them exactly how our galaxy looked in those early eons, with its long-gone quasars and all. They wouldn't see the sun and planets of our solar system, however, because they were not there at that time. They were not formed from the dust, gas, and debris of space until many billions of years later—probably only some 4 or 5 billion years ago today, as we learned earlier in Chapter 6.

THE FUTURE OF THE UNIVERSE

If we can learn how the universe began and how long ago it happened, can we also predict its future—how long it will be in existence, and what its end will be? Someday, perhaps—but not yet. There are two main possibilities. One is that the universe will continue to expand forever. If this happens, all the suns (stars) will burn out, one by one, as their fuel is exhausted, and the universe will eventually return permanently to the darkness of its early epochs. The burned-out cinders of stars, with their frozen, lifeless planets still orbiting around them, will fly away from each other in a forever-expanding and totally dark universe.

The other possibility is that gravitation, that is, the attractive force that each sun and each galaxy has for every other, will slow, and eventually stop, the expansion of the universe. Then everything will begin to move back, falling faster and faster under the pull of gravitation, toward the point of beginning. The whole universe would pile up at the fantastic speeds with which it departed the Big Bang explosion several hundred billion years before.

Although physicists admit that they cannot describe just what would happen, all life would certainly be snuffed out and everything destroyed in the flaming wreck of a "big crunch." In the enormous temperatures of the collapsing fireball, matter would be converted back into energy. Some people speculate that this fiery end could be another beginning—the energy released could produce another Big Bang that could begin the creation and expansion of another universe.

Cosmologists (scientists who deal with the origin, processes, and structure of the universe) are unsure whether the universe is "open" (will expand forever) or "closed" (will fall back to one point again in a "big crunch"). So far, there does not appear to be enough matter in the universe to produce the gravitational force needed to slow the expansion of the universe to a stop. However, measurements of the amount of matter are uncertain, and some cosmologists believe that there is much more matter out there in space than we know about. Someday, perhaps, we will learn whether the universe is headed for eternal darkness or an eternal cycle of explosion, expansion, contraction, explosion, expansion. . . .

The trees, the rocks, the earth, the sun, and the universe—all have clocks or calendars by which we have been able to find out how old they are. It was the search for the ages of things or the dating of events that led to a better understanding of what has happened during millions and billions of years to the world and to the universe around us. We now know much about when and how the human race began, when and how life on Earth began, and when and how the sun and the Earth were formed. We know something about when and how the universe began, but we don't know *why*. We may never know, possibly because there may be no answer to "why."

Appendix
RADIOACTIVITY—A QUICK REVIEW

Radiometric dating is based upon the natural process by which the atoms of one element (for example, uranium) change—or ''decay''—into atoms of another element (for example, lead). A general understanding of the structure of atoms will be helpful to understanding the dating process. This appendix presents a brief review of the basics of atomic structure.

THE MAKEUP OF THE ATOM

Although the atoms of different elements are different, all atoms are made up of the same basic particles. The central part, or nucleus, of an atom is always made up of one or more heavy particles called *nucleons*. Around this nucleus of heavy particles, other smaller, lighter particles, called *electrons,* move very rapidly in orbit. The simplest atom in the universe, hydrogen (Figure A-1), consists of one proton orbited by one electron.

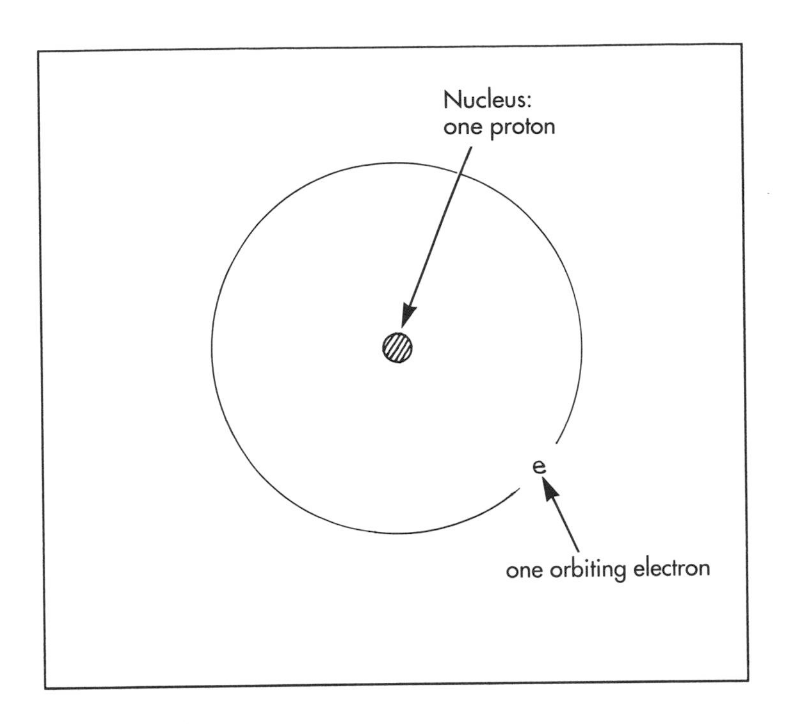

Figure A-1. The atom of the simplest element in the universe, hydrogen, consists of one proton orbited by one electron.

Protons • Some of the nucleons in each atom are positively charged particles called *protons*. The number of electrons orbiting the atom is normally equal to the number of protons in the nucleus. Because each electron has a negative charge equal and opposite to the positive charge of the proton, the net charge of an atom is ordinarily zero.

Neutrons • The remaining nucleons, which are found in atoms of all elements except hydrogen, are particles similar to protons, except that they have no electrical charge. These uncharged particles are called *neutrons*. For many elements, there are as many neutrons in the atomic nucleus as protons; for the heavier elements, even more. For example, helium has 2 protons and 2 neutrons in its nucleus, and 2 electrons in orbit around the nucleus (Figure A-2). Lead has 82 protons and 125 neutrons in its nucleus, with 82 electrons in orbit. Uranium, the heaviest natural element, is made up of 92 protons, 146 neutrons, and 92 electrons.

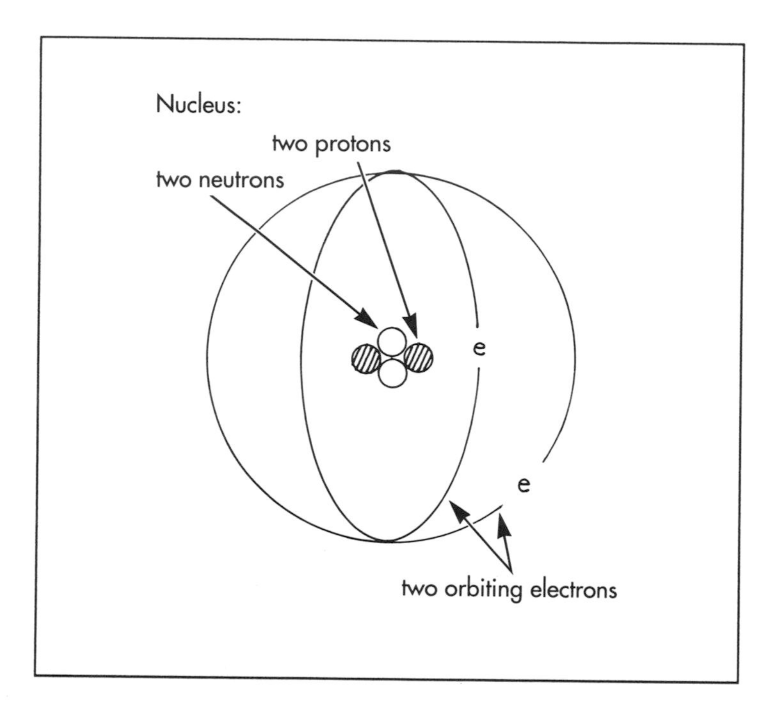

Figure A-2. The atom of helium consists of two protons, two neutrons, orbited by two electrons.

NUMBER OF PROTONS DETERMINES WHAT ELEMENT AN ATOM IS

To put it simply, it is the number of protons in an atom that determines what chemical element it is. An atom with 82 protons is always lead. One proton less (81) would make it an atom of the element thallium, while one proton more (83) would make it an atom of bismuth.

The number of neutrons in the nucleus, on the other hand, does not affect the chemical characteristics of the element. Atoms of a given element must have the required number of protons but may have a number of neutrons different from the normal number; such elements are called *isotopes.*

Isotopes • Isotopes are chemically identical to the basic element, though their extra neutrons cause them to have a different *mass.* For example, there are three isotopes of hydrogen, two of them rare compared to ordinary hydrogen. Both lead and uranium have several isotopes, with which we will be dealing shortly. In all, there are some 335 natural isotopes of natural elements.

Isotopes May Be Unstable • About sixty-five of these isotopes are "unstable," which means that they are spontaneously disintegrating through radioactive decay (some rapidly, others very slowly). All isotopes of uranium are unstable. Some isotopes of lead are unstable. If the atom gains or loses a proton in the process of decay, it will become an atom of a different element. If its number of protons does *not* change, it may, by gaining or losing a neutron, become just another isotope of the same element.

Chemical Processes Depend Upon Electrons • Strangely enough, although it is the proton content of the nucleus that determines the chemical nature of the element, the nucleus itself is not involved in chemical processes—those in which several kinds of atoms are combined to make a chemical compound. It is solely the *electron structure* that determines the chemical activity of an element and provides the bonding forces that hold atoms in chemical compounds. For example, the nuclei of atoms of hydrogen and the nuclei of atoms of oxygen are unchanged when the electron structure bonds the atoms of these two ordinarily gaseous elements together to form an entirely different material—the compound water (Figure A-3).

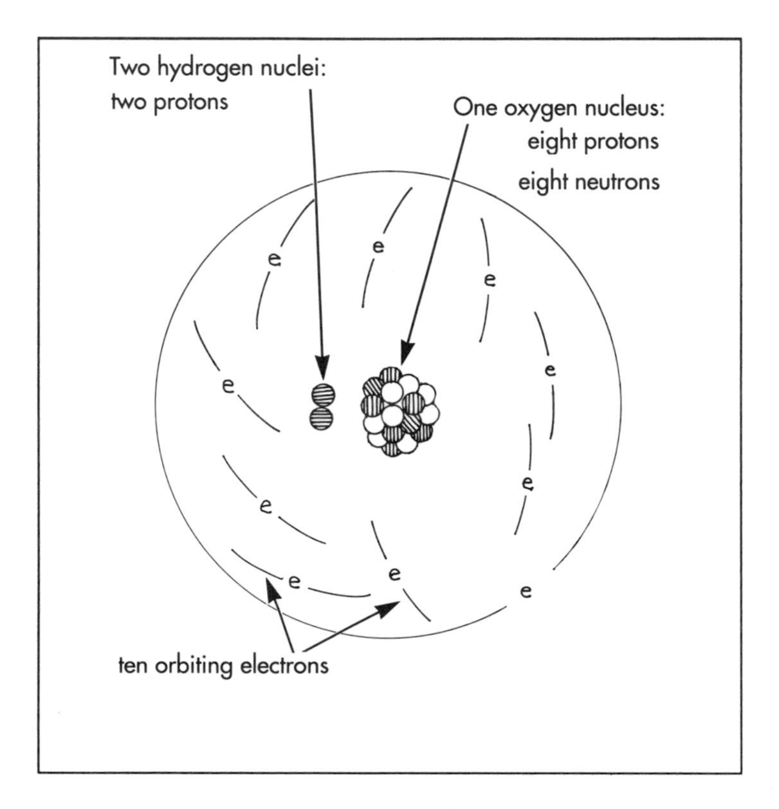

Figure A-3. An atom of the compound "water," chemical symbol H_2O, contains the nuclei of two atoms of hydrogen (two protons) plus the nucleus of one atom of oxygen (eight neutrons and eight protons) bound together by a cloud of ten orbiting electrons, one for each proton in the nucleus.

THE "DECAY" PROCESS

The process in which we are interested for use in radioactive dating is the natural fission process known as *radioactive decay.*

EJECTION OF PROTONS

The process of radioactive decay of the unstable element uranium, for example, begins with the spontaneous ejection of protons, which are nuclei of the helium atom. This ejection instantly changes the atom of ^{238}U into the atom ^{234}Th, an isotope of thorium that is itself unstable and will decay further into other elements.

ELECTRON EMISSION OR CAPTURE

Radioactive elements can also decay by electron capture or electron emission. When an electron is *captured* by an atomic nucleus, the positive charge of the capturing proton, in effect, is canceled and that proton becomes a neutron. Because it has one less proton, the element changes to the element on the periodic table that has one *less* proton. For example, an atom of potassium can decay into an atom of argon by this process.

When an electron is *emitted* from an unstable atomic nucleus, one neutron in effect loses its neutral charge to become a proton. The atom thereby changes to the element on the periodic table with one *more* proton. For example, an atom of radioactive hydrogen can decay into an atom of helium, or an atom of lead can decay into an atom of bismuth by emitting an electron. The decay of radioactive carbon, ^{14}C, into nitrogen, ^{14}N, occurs by this process and is the basis of another method of radioactive dating, as discussed in Chapter 7.

THE RADIOMETRIC CLOCK

The rate at which a radioactive (unstable) element decays into ''daughter products'' can be determined. This rate is unchanged by the conditions to which the element is exposed. A rock can be broken up, eroded away, exposed to heat or water, melted, or solidified, and the rate of decay of its radioactive atoms remains unchanged. This constant rate of decay provides a highly dependable ''clock'' that is the basis for radiometric dating. Measurement of the amount of daughter product in a sample, and application of the known decay rate, as described in Chapter 4, can tell us the amount of time during which the sample has been ''decaying,'' which is its age.

GLOSSARY

The definitions of words and phrases presented in this glossary pertain principally to their use in this book. More general definitions can be found in unabridged dictionaries. Only words that are widely used in the book are included. Words that are used only locally are generally defined there.

algae the earliest life on earth is believed to have been algae, primitive one-celled or multicelled plants with no true stems, roots, or leaves

artifact an object built or shaped by human workmanship, e.g., tools, pottery, weapons

asteroid the small planetary bodies and chunks of debris that orbit the sun between the orbits of the planets Mars and Jupiter.

brachiopod a kind of very ancient shellfish

Cambrian (period) the first period of the Paleozoic era, the era of ancient life. Taken from "Cambria," the Latin name for Wales, a principality of Great Britain, where deposits from this period were first found

Carboniferous a division of the Paleozoic era, characterized especially by the formation of coal deposits

Cenozoic (era) the most recent era of geologic time, characterized by the evolution of animals, birds, and plants

Cepheid a class of variable star with exceptionally regular periods of light pulsation

climatologist a scientist who studies climates

correlation (in geology) the process of identifying layers, or strata, of the earth's surface and determining their relationship over a large area using fossils or other geological tools

cosmic rays a stream of ionizing radiation coming from outside the earth

dating (in archaeology or paleontology) the art or science of determining the age of a rock, fossil, artifact, or other material

daughter product (also called "decay product") the element remaining after the occurrence of radioactive decay

decay (radioactive) the process by which an atom of an element expels a part of its nucleus and changes into an atom of another element

dendrochronology the study of growth rings in trees to determine and date past events (*dendra* from the Greek *dendron*—"tree"; *chrono* from the Greek *khronos*—"time"; *logy* from the Latin *logia*—"Science," "theory," or "study of")

deposition the act of depositing, e.g., in geology, the dropping of mud or sand by a stream

dike a mass of intruding igneous rock that cuts *across* the structure of adjacent rock

era (geologic) a particular division of geologic time, consisting of one or more geologic periods

erosion the breakdown and removal of rock or soil from any formation by dissolution, abrasion, and transportation

evolution (biology) the theory that organisms may, over generations, change physically from their ancestors

fossil the remains, trace, or imprint of an organism of past ages that is imbedded in the rocks or soil of the earth

galaxy a large-scale group of stars, gas, and dust, containing an average of 100 billion solar masses.

geologic column the sedimentary layers of the earth, arranged in a stack in the order of their deposition

geologist a scientist who works in the field of geology

geology the scientific study of the origin, history, and structure of the earth

glacier a huge mass of moving ice that has originated from compacted snow

Great Flood the worldwide deluge, survived by Noah and the Ark, as related in the Old Testament

half-life the time required for one-half of the radioactive atoms in a sample of radioactive material to decay into daughter products

Holocene (epoch) covers the postglacial period (or period of moderate glaciation) and the beginnings of human civilization to the present time

igneous (rock) formed by solidification from a molten or partially molten state

light-year the distance that light travels through the vacuum of space in the period of one year, approximately 5.9 trillion miles

Mesozoic (era) the third era of geologic time; the era which is characterized by the predominance of reptile life forms

metamorphic (rock) a rock mass that has been altered in composition, structure, or texture by great heat or pressure

meteorite chunks of matter that enter the earth's atmosphere from outer space and reach the surface of the earth

milky way the galaxy in which our solar system is located, visible as a hazy band of stars in the night sky

moraine an accumulation of boulders, stones, or other debris carried and deposited by a glacier

neutron an electrically neutral subatomic particle which, along with the proton, makes up the main mass of the atomic nucleus

paleontology the study of animal and plant fossils and the ancient life they represent

Paleozoic (era) the era of ancient life, beginning about 600 million years ago

periods (in geology) a unit of geologic time, shorter than an era and longer then an epoch

Permian (period) the last period of the Paleozoic era; a system of rocks or deposits from that period

Pleistocene (epoch) the geologic period during which northern glaciation occurred and the ancestors of the human race appeared

Precambrian (era) the oldest and longest era of geologic time, preceding the Cambrian

prism a transparent solid (such as glass) with triangular sides, used to break down a beam of light into its spectrum

quasar (astronomy) starlike objects that apparently have immense speeds, energies, and distances from Earth (from quasi-stellar, *quasi* meaning ''resembling, but not being'')

Quaternary (''fourth'') the period, starting at about 2½ million years ago, in which the human race developed

radioactivity the spontaneous emission of radiation from unstable atomic nuclei or from a nuclear reaction

radiometric (dating) a method of dating that is based upon radiation or radioactivity

radon a radioactive gas produced by the disintegration of the element radium

sedimentary (geology) pertaining to rocks formed from sediments, or from fragments transported and deposited in water

sill a sheet of igneous rock that has intruded, and lies *between,* layers of adjacent rock

spectrometer, spectroscope laboratory devices for producing, measuring, and analyzing spectra from a glowing material to determine its composition

spectrum, spectra the light pattern from a glowing material, produced by passing it through a prism so that it can be analyzed to reveal the composition or motion of the body

standard geologic column see geologic column

stratum (plural, **strata**) in geology, a layer of rock or soil having the same composition throughout

sunspots dark spots that appear in groups on the surface of the sun, believed to be storms or other disturbances in the sun's atmosphere

superposition (the principle of) the arrangement of strata in a definite sequence, with younger layers resting on top of older layers

tree rings (also growth rings) the circular arrangement of thin-walled cells and thick-walled cells by which a tree grows, one ring for each year; the rings are visible in any cut-off log or branch

trilobite a tiny ancient ancestor of the horseshoe crab

uniformitarianism the theory that all geological phenomena is explained by natural forces and processes that have operated from the formation of the earth to the present time

universe all existing things, including the earth, space, the galaxies, and all therein taken as a whole; the cosmos

varve a layer of sediment deposited in one year, as by a melting glacier

FOR FURTHER READING

A number of books are available on the various topics, such as the age of the earth, the history of geology, and the various kinds of dating. Many are useful to general audiences, others delve into specific topics to greater depth, while still others are highly technical textbooks. The following list covers some in the first two categories. No textbooks or highly technical titles are listed.

Annerino, John, *Hiking in the Grand Canyon.* San Francisco: Sierra Clubs, 1986.

Berry, William B. B. *Growth of a Historic Time Scale.* Cambridge, Massachusetts: Blackwell Scientific Publications, 1987.

Burchfield, Joe D. *Lord Kelvin and the Age of the Earth.* New York: Science History Publications, 1975.

Cooper, Henry S. *Moon Rocks.* New York: Dial Press, 1970.

Dalrymple, G. Brent. *The Age of the Earth.* Stanford, California: Stanford University Press, 1991.

Eicher, Don L. *Geologic Time*. Englewood Cliffs, New Jersey: Prentice-Hall, 1968.
Fenton, Carroll Lane. *Giants of Geology*. Garden City, New York: Doubleday, 1952.
Fishbein, Seymour L. *Grand Canyon Country*. Washington, D.C.: National Geographic Society, 1991.
Gould, Stephen Jay. *Time's Arrow, Time's Cycle*. Cambridge, Massachusetts: Harvard University Press, 1987.
Hurley, Patrick M. *How Old is the Earth?* New York: Anchor Books, Doubleday, 1959.
Moore, Ruth E. *The Earth We Live On: The Story of Geological Discovery*. New York: Knopf, 1956.
Porter, Roy. *The Making of Geology*. New York: Cambridge University Press, 1977.
Swinnerton, H. H. *The Earth Beneath Us*. Boston: Little Brown, 1955.
Wallace, Robert. *The Grand Canyon*. New York: Time-Life Books, 1972.
Zeuner, Frederick E. *Dating the Past, An Introduction to Geochronology*. London: Methuen, 1958.

INDEX

Italicized page numbers refer to illustrations.

Alpha Centauri (star), 98
Animal life, *40–41*, 91
extinct animals, 33, 37
fossil records of, 27, *28*, 29–30, 76–77, 84–85
Artifacts, 87, *88*, 89
dating of, 16, 22, 91–93, 95–96
Asteroids, 68–69
Atoms, 50, 89–90, 110–14

Becquerel, Antoine Henri, 49, 50
"Big Bang" theory, 103–4, 108
Bones, dating of, 92, 94
Brachiopods, 76, 84
Bristlecone pines, 14, *15*

Cambrian era, 75, 76–77, 84
Candles, time measurement by, 43–44, *44*
Carbon[14]. *See* Radiocarbon dating
Carbon dioxide, 90, 93
Carboniferous period, 77, 85
Catastrophe theory, 26–27, 30
Cenozoic era, 39, 77
Cepheid variable (star), 104–5, 106
Civilizations, 26
tree ring dating of, 14, 16, *17*, 18, *18*, 20
Climates, study of, 14, 16, 20–21
Coal deposits, 77, 85
Compounds, 50
Correlation, 38–39
Cosmic radiation, 68, 89, 93, 94
Creation theory, 25, 26, 27, 33, 37

Darwin, Charles R., 37
De Greer, Baron Gerard, 45
Doppler, Christian Johann, 102

Earth, age of, 23
ancient calculations of, 25–26
and catastrophe theory, 26–27, 30
and earth's crust, *24,* 27, 29, 30, 31
and fossils, 27, *28,* 29–30, 33, 36–39, 76–77, 84–85
and heat loss, 46–47
new measurements of, 74–75
and radioactive dating, 58, 63–64
Electrons, 110, 113–14, 115
Elements, 113
changes in, 51, 53
radioactive, 55–56, 60
Erosion and deposition, 31, 33, 62, 75, *79,* 82, 85
time measurement by, 42–43
See also Sediments
Evolution, 37, 38

Fault casts, 84
Fission-track dating, 60
Fluorine dating, 94–95
Fossils, 27, *28,* 29–30, 33, 36–39, 76–77, 84–85

Galaxies, 98, 100–1, 106
origin of, 103–4
Galileo, 100
Gamow, George, 103
Geological maps, 39
Geologic column, 39, *40–41,* 42, 63, 75–78, 83–84
Geologic periods of earth's history, 39, *40–41,* 42
methods of dating and, 42–43, *44,* 44–49
Glaciers, 45–46, 77–78
Grand Canyon, 78, *79,* 80, *81,* 82–86
Gravitational forces, 106, 108, 109
Great Flood, 26, 29–30

Half-life, 54, 55, 91
Heat loss, 46–47
Helium, 112, *112*
Holocene epoch, 78
Hubble, Edwin, 106
Human life, 33, 77, 78. *See also* Artifacts; Civilizations
Hutton, James, 31
Hydration, 96
Hydrogen, 60, 110, *111*

Ice Age, 77–78
Ions, 56
Isotopes, 55–56, 59, 60, 113

Lava, 63, 71, 72–73, 82
Lead, 51, *52, 53,* 54, *54,* 55, 56, 58, 59, 60, 112, 113
Libby, W. F., 89–90
Light, speed of, 97–98, 106

Light-years, 98
Lyell, Sir Charles, 31

Maria (seas) on the moon, 71, 72–73
Mass spectrometers, 56, *57,* 58, 59
Mesozoic era, 39, 77, 82, 85
Metamorphic rocks, 62, 63, 64, 82, 84
Meteorites, 65–66, *67,* 68–69, 73
Meteoroids, 66, 72
Moon craters, 69, *70,* 71
Moon rocks, 71–73
Moraines, 45

Natural selection, 37
Neutrons, 112, 113
Newton, Sir Isaac, 101
Nucleons, 110
Nucleus, 110

Obsidian, 95–96
Ocean salt, 48–49
Oil fields, 77

Paleozoic era, 39, 75, 76–77, 82, 84
Permian period, 77
Piltdown man, 95
Plant life, *40–41,* 91
  fossil records of, 29, 76–77, 84–85
Pleistocene epoch, 77, 95
Pottery, 87, 94
Precambrian era, 39, 76, 84
Protons, 111, 113, 115

Quasars, 105, 106, 107
Quaternary period, 77

Radioactive dating, 51, *52,* 53, *53,* 54, 89
  and earth's age, 58, 63–64
  of earth rocks, 58, 62–63
  elements used in, 55–56, 60
  mass spectrometer and, 56, *57,* 58, 59
  of meteorites, 65–66, *67,* 68–69, 73
  of moon rocks, 71–73
  problems in, 59–61
  reliability of, 61–62
Radioactive decay, 51, *52,* 53, *53,* 54, *54,* 113, 115–16
Radioactivity, 47, 49
Radiocarbon dating, 89–91, 115
  difficulties with, 92–93
  reliability of, 91–92
Radon, 59, 60
Rainfall, changes in, 21
Redshift, 102, 104–7
Rocks, dating of, 58, 62–64, 82, 84. *See also* Meteorites; Moon rocks

Rutherford, Ernest (Lord), 49, 50–51

Salt content of oceans, 48–49
Sedimentary rocks, 62–63, 64, 84
Sediments, 31, *32,* 34, 35, 45, 80, 82, 83, 84
Smith, William, 34, 36
Solar system, age of, 65, 66, 69, 73
Spectral analysis, 101
Starlight, 98, *99,* 100
analysis of, 101
measuring distances with, 104–5, 106
and star motion, 102–3
Stars, origin of, 103–4
Strata, *32,* 33, 34, *35,* 42, 75, 76, 78
fossils in, 36–39
in Grand Canyon, 78, *79,* 80, *81,* 82–86
superposition of, 34, 36, 76
See also Geologic column
Sunlight, speed of, 97–98
Sunspot cycles, 21
Superposition, 34, 36, 76

Temperature changes, 21, 72
Thermoluminescent dating, 94
Thompson, J. J., 56
Thomson, William (Lord Kelvin), 46–47
Tree ring chronology, 12, *12,* 13–14, *13, 15,* 89
climate study and, 14, 16, 20–21
dating of civilizations and, 14, 16, *17,* 18, *18,* 20
limits of, 20–22
Trilobites, *28,* 76, 84

Uniformitarianism, 31
Universe
early ideas of, 100–1
expansion of, 102–3, 104
future of, 108–9
measurement of, 106
origin of, 103–4
as a time machine, 97–98, *99,* 100, 107–8
Uranium, 47, 51, *52,* 53, *53,* 54, *54,* 55, 56, 58, 59–60, 64, 112, 113
Ussher, Bishop James, 25, 26, 74

Varves, patterns in, 45–46
Volcanic activity, 31, 63
dating of, 91–92, 94

Water, 114, *114*
Weathering processes, 83
Wien, Wilhelm, 56

ABOUT THE AUTHOR

Norman F. Smith was an aeronautical and aerospace research scientist with NASA—and its predecessor, NACA—for twenty-nine years. He was involved in wind-tunnel research in aerodynamics during the historic transition from subsonic to supersonic flight. He later held various engineering and executive positions during the Mercury, Gemini, and Apollo space projects.

After taking early retirement, Mr. Smith embarked on a new career as a science writer, lecturer, and educator. He has taught teachers, worked with school curricula, and produced textbook material, filmstrips, movies, and articles on a variety of science subjects.

"My particular interest is in getting good translations of the science behind interesting topics into the hands of students, teachers, and lay persons," Mr. Smith says. "My background in science permits me to tackle important, intriguing areas that have been neglected." *Millions and Billions of Years Ago* is one such topic, and Mr. Smith's seventeenth book for young people.